Dear *State of the World Library* Subscriber,

You hold in your hands the first of a newly redesigned *Worldwatch Papers* series. The Institute's original publication series, launched in 1975, has now been brought into a new era with a fresh design as well as other innovations: indexes, text boxes, and appendices. We expect that these changes will make the *Worldwatch Papers* more attractive, more readable, and more valuable as tools for research and advocacy.

What has not changed is our commitment to the solid interdisciplinary analysis and clear writing that have set the Institute's research and publishing apart since its founding.

We hope you find our redesigned *Worldwatch Papers* to be an improvement. Please let us know what you think of our new design—and how we might make the *Worldwatch Papers* even better.

Thank you too for your support of the Worldwatch Institute and our efforts to foster a sustainable world.

With optimism for a better future,

Christopher Flavin
President

Reading the Weathervane:

CLIMATE POLICY FROM RIO TO JOHANNESBURG

SETH DUNN

Jane A. Peterson, *Editor*

WORLDWATCH PAPER 160

August 2002

THE WORLDWATCH INSTITUTE is an independent, nonprofit environmental research organization in Washington, DC. Its mission is to foster a sustainable society in which human needs are met in ways that do not threaten the health of the natural environment or future generations. To this end, the Institute conducts interdisciplinary research on emerging global issues, the results of which are published and disseminated to decisionmakers and the media.

FINANCIAL SUPPORT for the Institute is provided by the Geraldine R. Dodge Foundation, the Ford Foundation, the Richard & Rhoda Goldman Fund, the William and Flora Hewlett Foundation, W. Alton Jones Foundation, Charles Stewart Mott Foundation, the Curtis and Edith Munson Foundation, David and Lucile Packard Foundation, John D. and Catherine T. MacArthur Foundation, Summit Foundation, Surdna Foundation, Inc., Turner Foundation, Inc., U.N. Environment Programme, U.N. Population Fund, Wallace Genetic Foundation, Wallace Global Fund, Weeden Foundation, and the Winslow Foundation. The Institute also receives financial support from its Council of Sponsors members—Tom and Cathy Crain, James and Deanna Dehlsen, Roger and Vicki Sant, Robert Wallace and Raisa Scriabine, and Eckart Wintzen—and from the many other friends of Worldwatch.

THE WORLDWATCH PAPERS provide in-depth, quantitative and qualitative analysis of the major issues affecting prospects for a sustainable society. The Papers are written by members of the Worldwatch Institute research staff and reviewed by experts in the field. Regularly published in five languages, they have been used as concise and authoritative references by governments, nongovernmental organizations, and educational institutions worldwide. For a partial list of available Papers, see back pages.

ISBN 1-878071-63-7

Library of Congress Control Number: 2002108840

Printed on paper that is 100 percent recycled, 100 percent post-consumer waste, process chlorine free.

BK
$7.00

Table of Contents

Figures, Tables, and Boxes

ACKNOWLEDGMENTS: I would like to thank Jonathan Pershing for providing his expertise in reviewing this paper. I owe a special note of gratitude to my editor, Jane Peterson, with whom I have had the pleasure of collaborating on four papers over the last five years. Thanks also to Lyle Rosbotham for the impressive cover redesign, to Gary Gardner for overall coordination, to Chris Flavin and Janet Sawin for comments and proofreading, and to Dick Bell, Susan Finkelpearl, and Leanne Mitchell for outreach. This paper is for Grace Ellen Drugge, my niece and goddaughter, for helping me to remember what is at stake.

SETH DUNN is a Research Associate with the Worldwatch Institute, where he has investigated energy and climate policy and strategy. He has participated in four rounds of the U.N. climate negotiations, including the 1997 Kyoto summit, and in expert meetings for the 2001 Third Assessment Report of the Intergovernmental Panel on Climate Change. He holds a BA in history and environmental studies from Yale University, and this fall will become a Senior Fellow with the Institute and a student at the Yale School of Management.

Introduction

Tens of thousands of citizens from government, industry, and civil society around the globe will gather in Johannesburg, South Africa, from August 26 to September 4, 2002, for the World Summit on Sustainable Development. More than 100 heads of state will mark the 10th anniversary of the historic Earth Summit, the 1992 U.N. Conference on Environment and Development in Rio de Janeiro, Brazil. The summit offers an opportunity not only to assess the world's progress in addressing the broad array of global challenges identified in Rio, but also to mobilize the political will needed for more meaningful action.[1]

One question certain to feature prominently at Johannesburg and beyond concerns the future of the Kyoto Protocol to the U.N. Framework Convention on Climate Change (UN FCCC). The original Framework Convention, which was agreed to in Rio and today has 186 member countries, established the objective of stabilizing atmospheric concentrations of greenhouse gases at levels that would avoid "dangerous anthropogenic interference with global climate," thereby minimizing adverse impacts such as rising sea levels, increased storm and flood frequency, the spread of infectious diseases, declines in biodiversity, and reduced availability of food and water. The Framework entered into force in 1994, and three years later, 181 nations meeting in Kyoto, Japan, agreed to the Kyoto Protocol, which would commit ratifying industrial nations to collectively reduce their emissions of these gases by 5 percent between 1990 and 2008–12.[2]

For the next three years, progress in finalizing the rules for implementing the Kyoto pact flagged. But new life was inadvertently breathed into the negotiations by the U.S. administration's withdrawal from the talks in early 2001, which galvanized Europe, Japan, Canada, and other industrial nations to resolve their points of contention. With the United States absent from the negotiating table, the remaining 180 nations worked out most of the outstanding issues at meetings in Bonn, Germany, and Marrakech, Morocco, in mid- and late-2001.[3] (See Box 1.)

As the Johannesburg Summit nears, whether or when the Kyoto Protocol will enter into force is fueling an international debate. This evokes a twinge of *déjà vu* when compared with the debate that preceded Rio, when the issue at hand was whether the original Framework treaty would be signed. Now, as then, the European Union (EU) leads the push to bring Kyoto into force, arguing that mandatory reductions in greenhouse gas emissions are essential. Now, as then, a U.S. administration led by a President George Bush resists European entreaties, citing concern that the terms under discussion might be unduly harmful to the U.S. economy. And now, as then, environmental NGOs furnish rosy studies projecting an economic boom if the agreement were to be ratified, while industry associations promote dire predictions of bust.

Beneath the surface of these superficial similarities, however, the environment of climate policymaking has undergone a sea change since 1992. At the time of the Rio summit, considerable scientific uncertainty existed as to whether human activities, rather than natural factors, could be conclusively implicated in ongoing changes in the Earth's climate. The costs of reducing greenhouse gas emissions were widely anticipated by economists to be very high, and the potential of new technologies to achieve substantial emissions cuts was only beginning to be calculated. For these reasons among others, most businesses were skeptical of the science of climate change and questioned the need for an international response. And the political clout of the United States, dampened by a recession and an unsympathetic administration, was sufficient to water down the initial framework agreement.

Ten years later, the landscape is fundamentally different. A broad scientific consensus now confirms that human-induced climate change is not only under way, but accelerating. Debate over the economics of climate change has broadened, in recognition of the ability of new technologies and policies to change conventional assumptions about the cost of reducing emissions. Wind and solar power, fuel cells and hydrogen, and other emerging energy technologies and fuels have entered or are nearing the marketplace, spawning nascent multibillion-dollar industries while dramatically improving prospects for a "decarbonized," low-greenhouse-gas energy economy. A growing number of corporations have moved beyond denial to acceptance of the science of climate change and have begun to act—seeking competitive advantage by anticipating future policy changes rather than resisting or reacting to them, in the belief that this approach will both reduce the long-run costs and increase the long-run benefits associated with action.

But the most significant development of all is that the international community may now have the multilateral momentum to bring the 1997 Kyoto Protocol into force—leaving the unilateralist United States standing on the sidelines. The European Union and its 15 member states, and Japan, gave the process a major boost with their ratifications in May and early June 2002. Should Russia and Canada follow suit—as they are expected to do—the conditions for the Protocol's entry into force would be satisfied.[4] (See Box 2.)

Amid these promising shifts in the political winds of change, it is helpful—and sobering—to remember that the Rio agreement, for all its initial promise, did little to keep global climate-related trends from heading in the wrong direction during the 1990s. Global carbon emissions set several new highs, while nearly all countries fell short of their initial Rio goals. New records in global carbon dioxide (CO_2) concentrations and global temperature were also established, suggesting that the gap between climate science and policy has widened, rather than narrowed, since Rio.

BOX 1

The U.N. Framework Convention on Climate Change and the Kyoto Protocol

The U.N. Framework Convention on Climate Change (UN FCCC), which was signed at the 1992 Rio Summit and entered into force in March 1994, established the objective of stabilizing atmospheric concentrations of greenhouse gases at levels that will avoid "dangerous anthropogenic interference with global climate" and allow economic development to proceed. The UN FCCC recognizes several basic principles: that scientific uncertainty must not be used to avoid precautionary action; that nations have "common but differentiated responsibilities"; and that industrial nations, with the greatest historical contribution to climate change, must take the lead in addressing the problem. The agreement commits all signatory nations to address climate change, adapt to its effects, and report on the actions they are taking to implement the Convention. It requires industrial countries and economies in transition—Annex I nations—to formulate and submit regular reports detailing their climate policies and their greenhouse gas inventories. And it requires Annex I nations to aim for a voluntary goal of returning emissions to 1990 levels by the year 2000, and to provide technical and financial assistance to developing, or non-Annex I, nations. Today, 181 nations and the European Union (EU) are party to the UN FCCC.

In 1995, signatories to the UN FCCC concluded that its existing commitments were inadequate for meeting its objective and launched talks toward a legally binding agreement that would establish emissions reduction commitments for industrial nations. These negotiations culminated in the 1997 Kyoto Protocol, which collectively commits industrial and former Eastern bloc nations—termed "Annex B" nations—to reduce their greenhouse gas emissions by 5.2 percent below 1990 levels during the period 2008-2012, and to demonstrate "meaningful progress" toward this goal by 2005. The agreement includes several measures designed to lessen the difficulty of meeting the target, such as the inclusion of six greenhouse gases and "flexibility mechanisms" that allow emissions trading, the use of forests and other carbon "sinks," and the earning of credits through overseas Clean Development Mechanism and joint implementation projects. It also commits all UN FCCC signatories to further advance their commitments to address greenhouse gas emissions.

In 1998, negotiators agreed to a plan of action and timeline for finalizing the rules surrounding the Protocol's specifics. At talks in the Hague in

Box 1 *(continued)*

late 2000, disagreement between the United States and the EU over several key provisions led to a breakdown in the talks. Following the U.S. withdrawal from the negotiating process in March 2001, a total of 180 nations finalized most of the "rulebook" for implementing the Protocol during talks in Bonn, Germany, and Marrakech, Morocco. Many of these details–placing no limit on trading, for example, or allowing credits from afforestation and reforestation–will allow countries more flexibility in meeting their Kyoto targets. The next round of negotiations will be held in New Delhi, India, from 23 October to 1 November 2002.

Source: See endnote 3.

At the same time, it would be a mistake to conclude—simply on the basis of these broader global trends—that climate policy has stood still, and that there is nothing to be learned from these last 10 years. In fact, hundreds of climate-related policies have been developed and adopted in many countries around the world. Understanding what has worked, what has not, and what has yet to be tried may be useful to countries now preparing to implement the Kyoto Protocol.

Despite its potential value, remarkably little effort has been devoted to evaluating the recent history of climate policy or to extracting its lessons, as this paper attempts to do. It begins by outlining how the environment of climate policymaking has changed over the past decade. It then analyzes and assesses the development and implementation of climate policy between Rio and Johannesburg in different countries, focusing on policies to reduce carbon emissions from the energy sector—the most important greenhouse gas and sector, respectively. It homes in on the experiences of 11 industrial and developing nations and one region—countries selected for their dominant influence on global carbon emissions trends (collectively, they put out more than two thirds of the global total), their prominent roles in international climate diplomacy, and their economic and geographic diversity.

The experience of this group is not only of historical interest—it carries momentous implications for climate policy.

BOX 2

How Can the Kyoto Protocol Enter into Force?

The Kyoto Protocol does not become an instrument of international law until it is brought into force. For this to occur, 55 parties, representing 55 percent of Annex I parties' carbon emissions in 1990, must ratify the agreement. As of early June 2002, the Protocol had been ratified by 74 parties. The majority of these are developing nations, including Argentina, Mexico, Senegal, and small island states such as Trinidad and Tobago. But the total also includes 21 Annex I parties, including the 15 member states of the EU, Japan, and several other European parties. With the U.S. unlikely to ratify in the near future, entry into force will require ratification by Russia and another party, possibly Canada or Poland.

Annex I Parties

Party	Share of Annex I Party 1990 Carbon Emissions (%)
U.S.	36.1
EU-15	24.2
Russia	17.4
Japan	8.5
Poland	3.0
Other Europe	5.2
Canada	3.3
Australia	2.1
New Zealand	0.2
Total	100.0

Annex I Parties that have ratified the Kyoto Protocol, June 2002

Party	Share of Annex I Party 1990 Carbon Emissions (%)
EU-15	24.2
Japan	8.5
Czech Republic	1.2
Romania	1.2
Slovakia	0.4
Norway	0.3
Iceland	0.0
Total	35.8

Source: See endnote 4.

Ten years ago, in Rio, it was widely believed that a voluntary aim—to return industrial-nation emissions to 1990 levels by 2000—would be a sufficient first step toward achieving the treaty's broader objective of climate stabilization. But with the benefit of a decade of hindsight, as the country studies that follow illustrate, it is evident that the entirely voluntary approach of the 1992 Rio framework treaty failed to motivate either effective domestic climate policy, or the international coordination of climate-related actions.

More specific lessons can be extracted from the early evolution of climate policy as well. The paper identifies several "good practices" that have been successful in areas such as fossil fuel subsidy reform and carbon taxes, energy efficiency, and support for renewable energy. It also points out "perverse practices" that have worked against the goal, most notably the tenacity of direct and indirect fossil fuel subsidies and the virtual neglect by most governments of the transport sector—the fastest-growing source of emissions.

Based on these case studies, it is clear that governments are experimenting with many climate policies, and that some are having a modest impact on emissions. But the aggregate impact of these efforts has been relatively slight, for several reasons: Many policies have been incompletely adopted, weak, or not fully implemented; many have not been coordinated with other measures or integrated with policies in other sectors; and few have been widely replicated. This indicates the need for governments to take a broader, more aggressive approach to climate policy.

This assessment also finds compelling evidence to dispel the notion promoted by Western opponents of climate change agreements that developing countries like China and India, which do not have binding commitments to limit emissions under the first round of Kyoto targets, are "rogue nation" emitters, undermining global climate goals. To the contrary, the study found developing nations, in advance of binding requirements to reduce emissions, already taking numerous actions to restrain emissions growth—largely for economic reasons. Their experience dramatizes the need for governments to better account for the "ancillary" economic benefits of cutting greenhouse gas emissions—such as reduced air pollution, energy cost savings, lower government expenditures, and the creation of new industries and jobs. The failure to factor in such benefits is but one of many flawed assumptions in the models of some economists, which project forbiddingly high costs from climate policy even as on-the-ground realities increasingly show such predictions to be highly exaggerated.

This decade of experience shows that the lack of industrial-country leadership in demonstrating the need to reduce emissions has been the largest obstacle to effective national and international climate policymaking. Recognition of this problem leads directly to the conclusion that the ratification and entry into force of the Kyoto Protocol would be the single most critical action needed to close the widening gap between climate science and climate policy—both internationally and nationally. Such a step would signal the renewed commitment by industrial-nation governments to the leadership role they signed up for in Rio, and prompt the private sector to prepare for accelerated development and market commercialization of clean energy technologies.

Bringing Kyoto into force would also be the best means of reengaging the U.S. government and private sector on this issue, and of setting the stage for discussing and phasing in fair, binding developing-nation commitments at a later stage. It would establish the global framework that is essential for ensuring that innovative and effective climate policy options—carbon taxation, emissions trading, overseas "clean development" projects—are internationally coordinated and widely implemented. And it would demonstrate, as one diplomat describing the recent revival of climate negotiations put it, "the triumph of multilateralism over unilateralism" in dealing with one of our most pressing global challenges.[5]

Troubling parallels exist between the current state of climate policy and the state of counterterrorism policy prior to the terrorist attacks on the United States on September 11, 2001. In both instances, what policies exist are mostly fragmented, reflecting a largely piecemeal approach and lack of coordination within and between government agencies. Even in the aggregate, these policies look somewhat Lilliputian next to the challenge, suggesting that national governments are greatly underestimating the scale and scope of the threat—and have failed to fully communicate to the public the degree of its vulnerability. These weak national commitments are both cause and effect of the fundamental problem: the absence of a strong, globally coordinated effort between governments. With both threats, the combination of national complacency and low global cooperation has resulted in a growing distance between the risks identified by experts and the responses made by policymakers.

In the case of terrorism policy, such deficiencies have been tragically exposed, and are belatedly being addressed through a multilateral, "global war on terror" and through national policy changes in many countries. One can only hope that a failure to revamp climate policy will not lead to the climatic equivalent of a terrorist attack, but—to borrow the parlance of military strategists and meteorologists alike—the possibility cannot be discounted. As the Johannesburg Summit approaches, what negotiators only tentatively recognized in Rio has become an unavoidable reality: that the most prudent insurance policy against climate change is global, multilateral action.

Advances in Science

The most important influence on climate policymaking is the authoritative work of the Intergovernmental Panel on Climate Change (IPCC). Established in 1988 by the U.N. Environment Programme (UNEP) and the World Meteorological Organization (WMO), the IPCC has a mandate to evaluate available scientific information on climate change and its potential impacts, and to devise a range of response strategies. Drawing on a network of hundreds of experts around the world for its periodic assessment reports, the panel meticulously collects, synthesizes, and reviews an enormous body of literature that spans the many fields relating to climate change.[6]

Each IPCC report has served as key input to the international climate negotiations. The 1990 assessment laid the foundation for negotiations that led to the 1992 Framework Convention. Similarly, the 1995 report provided the basis for negotiations that culminated in the 1997 Kyoto Protocol. And the latest review, published in 2001, has set the stage for the current round of negotiations over the Protocol's ratification, entry into force, and implementation. A brief overview of the 2001 assessment and comparison with its predecessors illustrate how the science, technology, and economics of the issue have evolved over the past decade.[7]

One important scientific debate during the 1990s centered on an observed increase in the global average surface temperature at the Earth's surface since the late nineteenth century. On the basis of several records, this increase is estimated to be in the range of 0.3–0.6 degrees Celsius. (See Figure 1.) At the time of the 1990 assessment, however, scientists were unable to determine for certain whether this change could be attributed to human influence, or whether the warming was due instead to natural variability, such as sunspots and volcanic eruptions.[8]

In subsequent years, however, scientists made considerable progress in distinguishing between natural and human influences. By accounting for the release of sulfate aerosols, which have a cooling effect on the atmosphere, they found a better match between computer-model simulations of climate change and the observed changes. This led the IPCC to assert in its 1995 report that "the observed warming trend is unlikely to be entirely natural in origin" and that "the balance of evidence suggests a discernible human influence on global climate."[9]

In the years following the release of the second assessment, new studies of past and current climates and better analysis and comparison of data sets further advanced scientific understanding of climate change. In the third assessment report, in 2001, the IPCC found that "an increasing body of observations gives a collective picture of a warming world and other changes in the climate system." The observed changes include widespread decreases in snow cover and ice extent and a rise in sea level of 10–20 centimeters during the twentieth century. The panel concluded that the 1990s were likely the warmest decade since instrumental record-keeping began in the late nineteenth century, with 1998 being the warmest year. According to measurements in the northern hemisphere, the average global surface temperature rose more during the twentieth century than during any other century in the last 1,000 years.[10]

FIGURE 1

Global Average Surface Temperature, 1880–2001

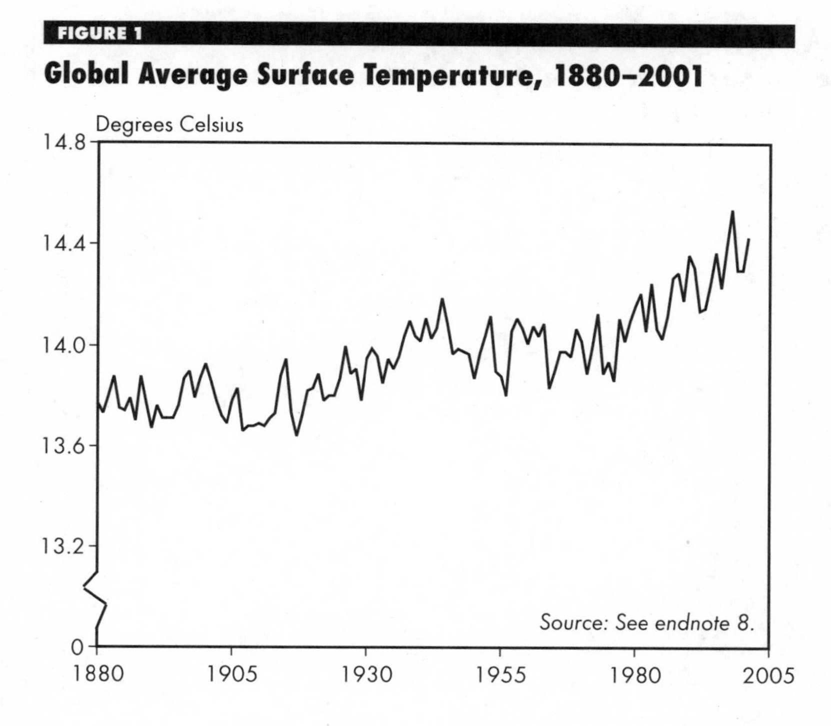

This rise in global temperature has occurred in tandem with rising levels of greenhouse gas. Since 1750, atmospheric concentrations of CO_2, the most significant greenhouse gas, have risen by 31 percent, with more than half of this increase occurring in the last 50 years. (See Figure 2.) Current CO_2 concentrations are at their highest in the last 420,000 years, according to ice core data; studies of plankton suggest they may be at their highest in the last 20 million years. Based on this evidence, the IPCC concluded that "there is new and stronger evidence that most of the warming observed over the last 50 years is attributable to human activities."[11]

CO_2 is not the only greenhouse gas contributing to this warming. Other important gases, emitted primarily from agricultural and industrial practices, include methane, nitrous oxide, sulfur hexafluoride, hydrofluorocarbons, and perfluorocarbons. But CO_2 is the most important gas: It accounts for more than half of the human-induced warming that has

FIGURE 2

Atmospheric Concentrations of Carbon Dioxide, 1958–2001

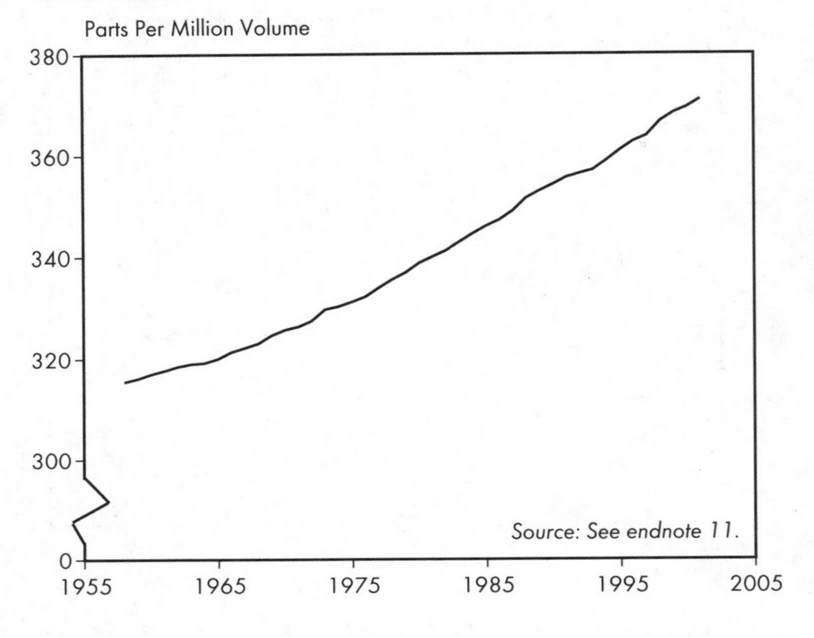

already occurred, and is projected to account for three fourths of the warming that will have taken place by 2100. And of the carbon emissions of the past 20 years, those from fossil fuel burning comprise three fourths, with deforestation responsible for the remainder.[12]

The IPCC thus identifies fossil fuel combustion as the dominant influence on future trends in global carbon emissions, which currently stand at approximately 6.2 billion tons per year. Under the IPCC scenarios, rising carbon emissions are projected to raise CO_2 concentrations to between 540 and 970 parts per million volume (ppmv) by 2100. These levels correspond to roughly more than a doubling and tripling, respectively, of pre-industrial CO_2 concentrations.[13]

Corresponding to the higher CO_2 levels are increases in global average surface temperature. The 2001 assessment projects an increase of 1.4–5.8 degrees Celsius between 1990 and

2100. This represents a wider range than the 1.5–4.5-degree estimate of the 1995 assessment—not because of greater uncertainties, but because a broader set of assumptions and scenarios were considered. In any case, the rate of temperature change implied by these projections would be much higher than those of the past century, and probably unprecedented in the last 10,000 years. Average sea level is projected to rise by 9–88 centimeters. Also projected are a continued shrinkage in snow cover and sea ice, and a more widespread retreat of glaciers and ice caps. Even if greenhouse gas concentrations are stabilized, climate change will persist for many centuries, with surface temperature and sea level continuing to rise in response to past emissions.[14]

The 2001 assessment also reveals greater confidence in the expected impacts of, and vulnerability to, climate change. Higher minimum and maximum temperatures, higher heat indices over land masses, and more intense precipitation events were all "likely" or "very likely" to have occurred over the past century—and are "very likely" to be seen in the future. There is stronger evidence of impacts on biological and physical systems, including glacier shrinkage, permafrost thawing, and declines of plant and animal populations. Several natural systems are recognized as especially at risk of irreversible damage, including glaciers, coral reefs and atolls, boreal and tropical forests, and polar and alpine ecosystems. Climate change is also expected to exacerbate the risks of extinction of the more vulnerable species and the loss of biodiversity.[15]

Many human systems are vulnerable to climate change: water resources, agriculture, forestry, coastal zones and marine systems, human settlements, energy, industry, insurance and financial services, and human health. Some of the projected adverse effects include:

- a reduction in potential crop yields in most tropical and subtropical regions for most temperature increases;
- decreased water availability for populations in many water-scarce regions, most notably in the subtropics;

- an increase in the number of people exposed to vector-borne and water-borne diseases (such as malaria and cholera) and a rise in heat stress mortality; and
- a widespread increase in the risk of flooding for tens of millions of people, due to both increased heavy precipitation events and sea level rise.

In addition, projected changes in climate extremes—droughts, floods, heat waves, avalanches, and windstorms—could have major consequences, as the frequency and severity of these events are expected to increase.[16]

The latest IPCC assessment also reflects greater attention to large-scale, irreversible impacts, which pose risks that have not been reliably quantified. Examples include a significant slowing of the ocean circulation system that conveys warm water to the North Atlantic, major reductions in the Greenland and West Antarctic ice sheets, accelerated warming due to carbon releases from terrestrial ecosystems, and the release of carbon from permafrost regions and methane from hydrates in coastal sediments. If these changes occurred, their impact would be widespread and sustained. Slowing of the oceanic circulation would reduce warming over parts of Europe. Loss of either the West Antarctic or the Greenland Ice Sheet could raise global sea level up to three meters over the next 1,000 years, which would submerge many islands and inundate extensive coastal areas. Added carbon and methane releases would further amplify the warming.[17]

Scientists have also devoted more study to ways in which countries can adapt to climate change. But adaptation can entail significant costs, and the most vulnerable countries have the fewest resources and the least ability to adapt. The IPCC concluded that "the effects of climate change are expected to be greatest in developing countries in terms of loss of life and relative effects on investment and the economy." Regional assessments reveal major vulnerabilities around the globe—a National Research Council study points to serious adverse impacts in the United States—but the countries hit hardest are the ones that have contributed least to the problem.[18]

New Views on Solutions

How much climate change occurs will depend primarily on how high CO_2 concentrations rise, which in turn will be determined by trends in carbon emissions from fossil fuel burning. Stabilizing greenhouse gas concentrations at 450 ppmv, for example, requires that annual carbon emissions drop well below current levels within the next several decades, then to around 2 billion tons by 2100, and ultimately to less than 1 billion tons. This entails a cut of roughly 70–80 percent in global carbon emissions—much larger than the current Kyoto cuts.[19]

Lowering global carbon emissions will require major changes in existing patterns of energy resource development. Fortunately, the potential of new technologies and policies to slow climate change has grown dramatically since Rio. Since its 1995 assessment, reports the IPCC, "significant progress relevant to greenhouse gas emissions reduction has been made and has been faster than anticipated." Advances are taking place in a wide range of technologies now in varying stages of development. These include the commercial introduction of wind turbines, the elimination of industrial byproduct gases, the emergence of highly efficient hybrid-electric cars, and the advance of fuel cell technology toward commercial viability.[20]

What is the potential for lowering emissions in the relatively near future? Summarizing hundreds of studies, the IPCC concludes that global emissions could be reduced well below 2000 levels between 2010 and 2020. Specifically, the panel estimates that emissions could be reduced by 1.9–2.6 billion tons of carbon equivalent by 2010, and then by 3.6–5.5 billion tons by 2020. (At the moment, emissions are projected to reach 11.5–14 billion tons by 2010 and 12–16 billion tons by 2020.) The panel also found that at least half of these reductions could be achieved by 2020 in a cost-effective fashion, or at no net cost.[21]

The low-cost opportunities lie primarily in the hundreds of technologies and practices that promote efficient energy use in buildings, transportation, and manufacturing. In addition, natural gas is expected to play an important role in reducing emissions in tandem with power plant efficiency improvements and greater use of cogeneration (the combined use of heat and power). Important contributions can also be made by low-carbon energy systems, such as biomass from forestry and agricultural byproducts, landfill methane, wind and solar power, hydropower, and other renewable sources of energy. Agriculture and industry can reduce other greenhouse gases: Methane and nitrous oxide emissions can be cut from livestock fermentation, rice paddies, nitrogen fertilizer use, and animal wastes, while process changes and the use of alternative compounds can minimize the emissions of fluorinated gases.[22]

Using these available or near-ready technologies, most models suggest that atmospheric CO_2 levels could be stabilized at 450–550 ppmv, if not lower, over the next 100 years. Bringing this about, however, would require major socioeconomic and institutional changes. It would also imply an accelerated reduction of the carbon intensity of the global economy, as measured by emissions per unit of output. It further suggests that energy supplies could no longer be dominated by low-priced fossil fuels.[23]

What are the costs and benefits of cutting emissions? Analyses vary widely, depending on which methodologies and underlying assumptions are applied. Estimates depend, for example, on whether the revenue of carbon taxes is recycled back into the economy through reductions in other taxes; whether the benefits of avoided climate change—including side benefits such as energy savings, reduced local and regional air pollution, energy security, and employment—are factored in; and whether the external costs of climate change are incorporated into market prices. Other assumptions shaping models of the economics of climate change include demographic, economic, and technological trends; the level and timing of the agreed-to target; and the degree of reliance on various implementation measures, such as emissions trading.[24]

There is a consensus among experts that some greenhouse gas emissions can be limited at no cost, or even at a net benefit to society through "no regrets" policies—which would make economic sense even if climate change were not an issue—that address imperfections in the market. A lack of information, for instance, can prevent consumers and businesses from adopting efficient technologies that lower overall energy costs. If carbon taxes or auctioned emissions permits are used to offset reduced wage and labor taxes, the benefits can increase. In many cases, the ancillary benefits of reducing carbon emissions—lower air pollution, new jobs, reduced oil imports—balance out the costs dictated by the policies themselves. Reducing carbon emissions can also lower emissions of particulates, ozone, and nitrogen and sulfur oxides—thus creating significant human health benefits such as pushing back the incidence and severity of asthma and other respiratory problems.[25]

Some greenhouse gas emissions can be limited at no cost, or even at a net benefit to society.

Recent government studies support the notion that there is significant potential for low-cost or no-cost emissions cuts. A U.S. Department of Energy study, for instance, estimates that the nation could achieve most of its Kyoto cuts at no net cost, primarily by removing market barriers to the adoption of existing energy-efficient and renewable energy technologies. These policies would also cut back air pollution, petroleum dependence, and inefficiencies in energy use, leading to economic benefits that counterbalance overall costs. Similarly, a report from the Climate Change Programme of the European Commission indicates that the European Union can reach its Kyoto target through cost-effective measures that amount to no more than $18 per ton of carbon dioxide, accounting for about 0.6 percent of the region's gross domestic product (GDP). These measures, primarily involving enhanced energy efficiency, have the potential to achieve more than double the emissions cut required of the EU under the protocol.[26]

How much would it cost industrial and former Eastern bloc nations (Annex B countries) to implement the Kyoto

Protocol? That depends on how much trading of emissions is involved and what domestic measures are taken. Without emissions trading between these countries, most global studies show reductions in projected GDP of about 0.2–2 percent in 2010 for different regions. With full emissions trading, however, the reductions would be just 0.1–1.1 percent of projected GDP—amounts so small that they would likely disappear in the "noise" of natural variations of the economy. The models also do not factor in the use of carbon sinks or non-CO_2 greenhouse gases to meet targets, the Clean Development Mechanism (See Box 1), side benefits, or revenue recycling.[27]

Economies in transition, which are included in the Annex B grouping, represent a special case. For most of them, the projected effects of implementing Kyoto range from an increase of several percent of GDP to negligible—reflecting enormous opportunities for improving energy efficiency. If energy efficiency is indeed improved drastically, emissions in 2010 could be well below the amounts assigned to them under the Kyoto treaty. In such instances, models show a rise in GDP, due to revenues these countries obtain from selling their emissions trading surpluses—the amount left over after they have met their targets.[28]

What would it cost to reduce emissions more aggressively? Conventional economic models typically suggest that costs will rise as the level at which greenhouse gas concentrations are stabilized drops (from 750 to 550 ppmv, or from 550 to 450 ppmv). But these models ignore the potential of more ambitious targets to bring about deeper technological change by spurring industry to make large, rather than incremental, innovations. "Induced technological change" is an emerging field of research in climate change economics, but most models do not account for this phenomenon. Those models that do incorporate the effect of policies on technological change suggest that a number of policy packages could lead to a stabilization of CO_2 concentrations while allowing continued GDP growth.[29]

Efforts to improve climate-related modeling have resulted

in the "integrated assessment" model, which attempts to synthesize climate science, policy, and economic research. This type of model is useful for assessing policies, coordinating issues, and comparing climate and non-climate policies and it is becoming increasingly influential in policy circles. But a study from the Pew Center on Global Climate Change observes that most integrated assessment models are based on economic theories with simplifications that do not always apply to climate policy. In particular, they make unrealistic assumptions about how market forces drive technological innovation, the behavior of firms, intergenerational equity, and climate "surprises." Such misguided assumptions tend to drive up the estimated cost of dealing with climate change.[30]

However the costs and benefits add up, they will be spread unevenly among different sectors of the economy. Generally speaking, it is easier to identify the sectors that are likely to face economic costs than it is to pinpoint those that may benefit. In addition, the costs are more immediate, more concentrated, and more certain, even if the benefits prove to be greater. Coal and certain energy-intensive sectors—such as steel production—are most likely to suffer an economic disadvantage. Others, including the renewable energy industry, are expected to benefit over the long term from price changes and the availability of financial and other resources that might otherwise have been committed to carbon-intensive energy sectors.[31]

Appropriate measures can help cushion some of the costs to various sectors. The removal of fossil fuel subsidies would enhance total societal benefits by improving economic efficiency, while emissions trading will cut the net economic cost of meeting the targets. Some policies, such as exempting energy-intensive industries from carbon taxes, will redistribute the costs but also increase the total expense to society. And the revenues from a carbon tax can be used to compensate low-income groups who would otherwise suffer.[32]

Countries that do not have initial emissions constraints will be indirectly impacted by those that do. For oil-exporting developing countries, some studies estimate impacts as high as a 13 to 25 percent reduction of projected oil revenues by

2010. But these studies do not consider policies other than trading—which could lower the impact on oil exporters, and thus tend to overstate both the costs to these countries and the overall costs. Such nations can further reduce the impact by removing subsidies for fossil fuels (thus lowering government expenditures), restructuring energy taxes according to carbon content, increasing natural gas use, and diversifying their economies.[33]

Other developing countries can expect both costs and benefits. They may suffer the effects of reduced demand for exports and the higher price of imports. At the same time, however, they may benefit from the transfer of environmentally sound technologies and know-how. Because no country is likely to experience the same net effect as another, it is hard to identify winners and losers. As for "carbon leakage"—caused by carbon-intensive industries relocating to developing countries in response to changing prices—the estimates range from a 5 to a 20 percent increase in these countries. But these models do not account for the transfer of environmentally sound technologies and skills, which could lower—and in the longer term more than offset—the environmental or economic costs of any leakage.[34]

Climate Policy Report Card

Even if numerous technological and economic opportunities exist for reducing greenhouse gas emissions, governments need to overcome the many technical, economic, political, social, behavioral, and institutional barriers that typically keep these opportunities from being fully exploited. The options vary substantially by region and sector and over time, with poor people facing particularly limited options for adopting technologies or changing behavior. In industrial countries, the opportunities relate primarily to social and behavioral barriers; in economies in transition, they center on rationalizing prices; in developing countries, they hinge largely on greater access to information and advanced technologies, financial resources, and training. To be sure, every country can find opportunities to surmount some combination of these barriers.[35]

One constant theme in the existing literature on climate policy is that national responses to climate change can be more effective if they are deployed as a *portfolio of policy instruments* that either limit or reduce greenhouse gas emissions. These instruments might include:

- carbon/energy taxes,
- tradeable permits,
- removal of subsidies to carbon energy sources and provision of subsidies to carbon-free sources,
- refund systems,
- technology or performance standards,
- energy mix requirements,
- product bans,
- voluntary agreements,
- government investment, and
- research and development investment.

While there is no one policy of choice, policy experts increasingly promote market-based instruments as being especially cost-effective. Studies suggest that the removal of fossil fuel subsidies can result in "win-win" situations, improving efficiency and reducing environmental damage. Estimates of the cost of global price subsidies to fossil fuel consumers range from $215 billion to $235 billion; World Bank economists calculate that removing these distortions would cut annual carbon emissions by 1.4 billion tons by 2010. According to the Organisation for Economic Co-operation and Development (OECD), a combination of fossil fuel subsidy removal and carbon taxes would cut carbon emissions in OECD member countries by 15 percent between 1995 and 2020. The IEA estimates that removing fossil fuel subsidies in China, Russia, India, Indonesia, Iran, South Africa, Venezuela, and Kazakhstan would reduce carbon emissions in these countries by 16 percent—and global emissions by 4.6 percent.[36]

Also recommended are energy efficiency standards, which were already widely used prior to the advent of climate policy,

FIGURE 3

Global Carbon Emissions from Fossil Fuel Burning, 1950–2001

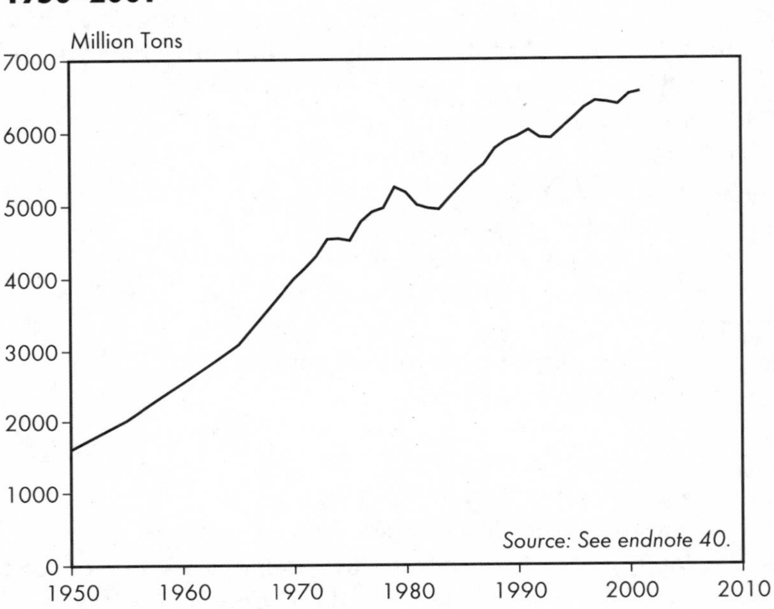

and could be effective in a number of countries. Voluntary agreements with industry have become more frequently relied upon in many areas of environmental policy, in some instances as a precursor to more stringent measures. Other possible measures include influencing consumer and producer behavior through information campaigns, environmental labeling, green marketing, and incentives. And government and private R&D are essential for advances in technologies that will lower costs further.[37]

Another common observation from experts is that climate policy can be more effective when *integrated with the "non-climate objectives" of national and sectoral policies* and translated into broader strategies for long-term technological and social change aimed at sustainable development. Just as climate policies indirectly achieve socioeconomic benefits, non-climate policies can yield climate benefits. For example, emissions

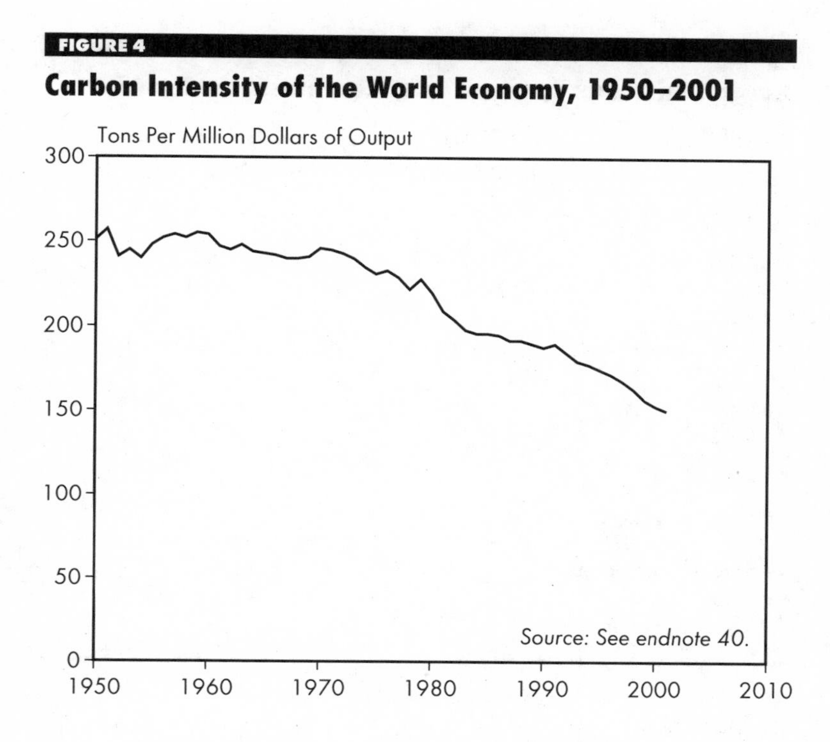

could be reduced significantly through socioeconomic policies such as energy infrastructure development, pricing, and tax policies. Transferring climate-friendly technologies to small- and medium-sized enterprises is another case in point. Acknowledging these *ancillary benefits* can also convince leaders to lower the political and institutional barriers to actions pertaining to climate.[38]

Coordinating actions is another way to reduce costs and can also avoid conflicts with international trade. Taxes, standards, and subsidy removal can also be coordinated or harmonized, though steps to do so have so far been limited. As for the *timing of policies*, the IPCC has reaffirmed the finding of its 1995 report: The earlier action is taken to mitigate climate change, the greater flexibility there will be in moving toward the stabilization of atmospheric greenhouse gas levels. Economic models completed since the second assessment suggest that a gradual transition toward a less carbon-emitting energy system

TABLE 1

Kyoto Emissions Targets, First Commitment Period

Country/Region	Target[1]	Actual Emissions[2]
	1990-2008/12	1990-2001
	(percent)	(percent)
United States	-7	+15.7
European Union	-8	-0.2
Germany	-21	-17.1
United Kingdom	-12.5	-4.1
Japan	-6	+10.8
Canada	-6	+ 11.5
Australia	+8	+32.3
Russia	0	-30.5
All Annex B countries	-5.2	-1.7

[1] All greenhouse gases.
[2] Carbon dioxide only.
Source: See endnote 41.

would minimize the premature retirement of power plants, factories, and other forms of capital stock. It would provide time for technology development and avoid a premature lock-in to early versions of low-emission technologies that are developing rapidly. And greater near-term action would give needed breathing room, decreasing the environmental and human risks associated with rapid climate changes, allowing for a later tightening of targets, and addressing concerns about the effectiveness and equity of the climate regime.[39]

Despite the strengthening case for climate policy, the record of the past decade has been mixed. Global carbon emissions from fossil fuel combustion rose by 10.2 percent between 1990 and 2001. (See Figure 3.) The cumulative global carbon emissions between 1990 and 2001, over 74 billion tons, reflect a 14 percent increase over the 65 billion tons emitted worldwide between 1980 and 1991. The transport sector was the fastest-growing source of carbon emissions in the 1990s, recording a torrid 23 percent rise and increasing its

TABLE 2

Climate Policy Report Card, Countries/Regions Studied

Country	2001 Emissions	Rank	Global Emissions Share	2001 Population	Global Population Share	Per Capita Emissions
	(million tons)		(percent)	(millions)	(percent)	(tons/person)
Australia	95	14	1.4	19	0.3	5.0
Brazil	82	17	1.2	172	2.8	0.5
Canada	129	9	2.0	31	0.5	4.2
China	721	3	11.0	1,273	20.7	0.6
European Union	829	2	12.7	377	6.1	2.2
Germany	218	7	3.3	82	1.3	2.7
United Kingdom	147	8	2.2	60	1.0	2.5
India	304	6	4.6	1,033	16.8	0.3
Japan	311	5	4.8	127	2.1	2.5
Russia	403	4	6.2	144	2.4	2.8
South Africa	91	15	1.4	44	0.7	2.1
United States	1,510	1	23.1	285	4.7	5.3
Total	4,475	–	68.4	3,505	57.1	–
Global average	–	–	–	–	–	1.1

Source: See endnote 42.

overall share to 24 percent. Emissions from energy production grew by 22 percent, and account for 43 percent of emissions; emissions from industry dropped by 10 percent and have a 19 percent share; and emissions from the residential and commercial sectors fell by 8 percent, and account for 14 percent of overall emissions. The carbon intensity of the global economy fell by 20.1 percent, due to increased use of natural gas, declining use of coal, overall improvements in energy efficiency, and structural shifts to less energy-intensive sectors.[40] (See Figure 4.)

As for the Kyoto Protocol's commitment of industrial and former Eastern bloc countries to reduce greenhouse gas emissions by 5.2 percent between 1990 and 2008–12, this group of nations reduced carbon emissions by 1.7 percent between 1990 and 2001. (See Table 1.) In non-Annex B countries, carbon emissions rose by 29.9 percent. Annex B countries still account for the majority—58 percent—of global carbon emissions.[41]

The following sections describe, in more detail, the carbon emissions trends and climate policies of 11 countries and one region listed alphabetically. This group collectively accounts for 68 percent of global carbon emissions and 57 percent of global population and encompasses most of the world's top 20 emitters on both an absolute and per-capita basis. (See Table 2.) While they have widely varying commitments under the UN Framework Convention on Climate Change and the Kyoto Protocol, their experiences yield some valuable lessons.[42]

Australia

Australia accounts for 1.4 percent of global carbon emissions and ranks 14th globally. Its per capita emissions, at 5 tons per person, are the world's second highest and more than four times the global average. Under the Kyoto Protocol, Australia committed to limiting emissions growth to 8 percent between 1990 and 2008–12. Through the year 2001, however, Australian carbon emissions had climbed to 32.3 percent above 1990 levels.[43] (See Figure 5.)

This trend was driven by increases in consumption of coal, oil, and natural gas, by 20.5, 20.6, and 23 percent, respectively. Australia plans to achieve a relatively large portion of its Kyoto target through carbon sequestration in its rangeland and forests, but continued carbon emissions growth would impede efforts to reach its Kyoto goal. Australia's carbon intensity fell by 9.2 percent, but remains among the world's highest.[44] (See Figure 6.)

In 1998, the Australian government approved a National Greenhouse Strategy aimed at lowering greenhouse gas emissions from 43 percent above 1990 levels—the business-as-usual path—to 18 percent above 1990 levels by 2010. It established the Australian Greenhouse Office for coordinating policy, set a mandatory target for electricity retailers to derive an additional 2 percent of their electricity from renewable energy by 2010, and set aside funding for commercializing renewable energy technologies. Other measures include accelerating market reform and instituting efficiency standards in power generation, improving motor vehicle efficiency by 15 percent, and adopting building codes and standards.[45]

Although Australian government experts maintain that climate change is being given a high priority and that they are targeting ways to reduce emissions more systematically, outside analysts are less certain. A 1999 expert review of Australian policy reported that "it is difficult to accurately assess Australia's progress toward meeting its greenhouse gas (GHG) target of reducing GHG emissions to 108 percent of 1990 levels by the year 2010, since many of the Government's policies and measures are still in the planning or early implementation phase and baseline information on carbon sink status related to land use change is still being refined." While some measures under consideration have since been adopted, their impact on recent emissions trends has been limited.[46]

Alternative fuel use in transport, voluntary agreements with industry, and increased support for renewable energy research and application are the main parts of Australian

FIGURE 5

Carbon Emissions, Australia, 1990–2001

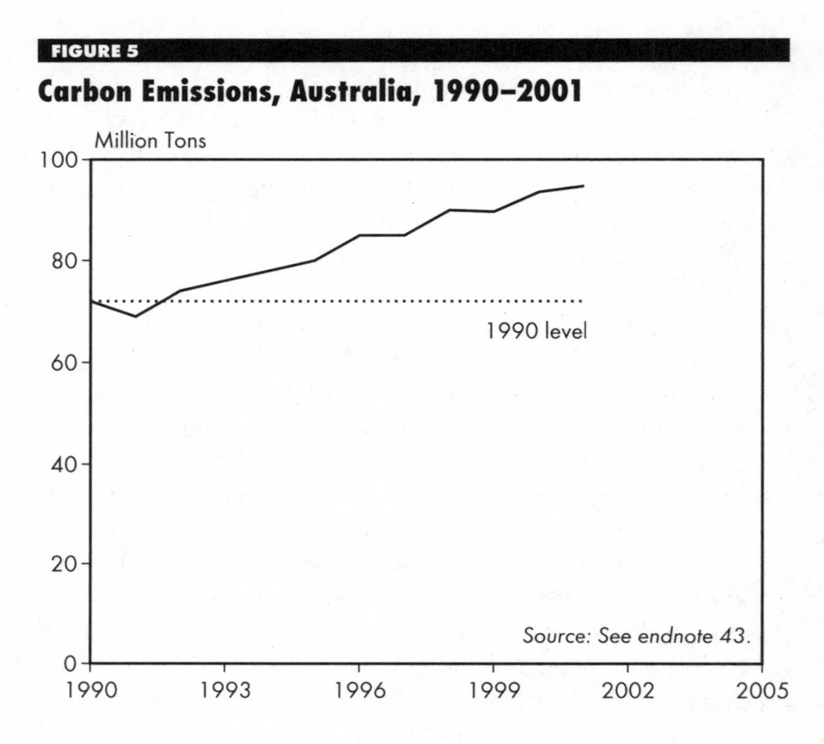

energy-related climate policy. There is no national system in place to monitor Australian climate policy across sectors, although some programs have been periodically reviewed. According to an independent review team, some 70 percent of the expected reductions are dependent on cooperative agreements with industry that emphasize voluntary, cost-effective, or "no regrets,"activities. There are few regulatory initiatives, and no measures that employ fiscal instruments.[47]

The Greenhouse Challenge Program (GCP) focuses on voluntary reduction agreements with individual firms. While some successes have been reported under this approach, Australian experts concluded that voluntary action alone would not be sufficient to bring emissions down to the level of the national target. This led to the announcement of new cost-effective initiatives for energy-efficient power generation, transportation efficiency, and building codes.[48]

The GCP covers industries comprising about 45 percent

FIGURE 6

Carbon Intensity, Australia, 1990–2001

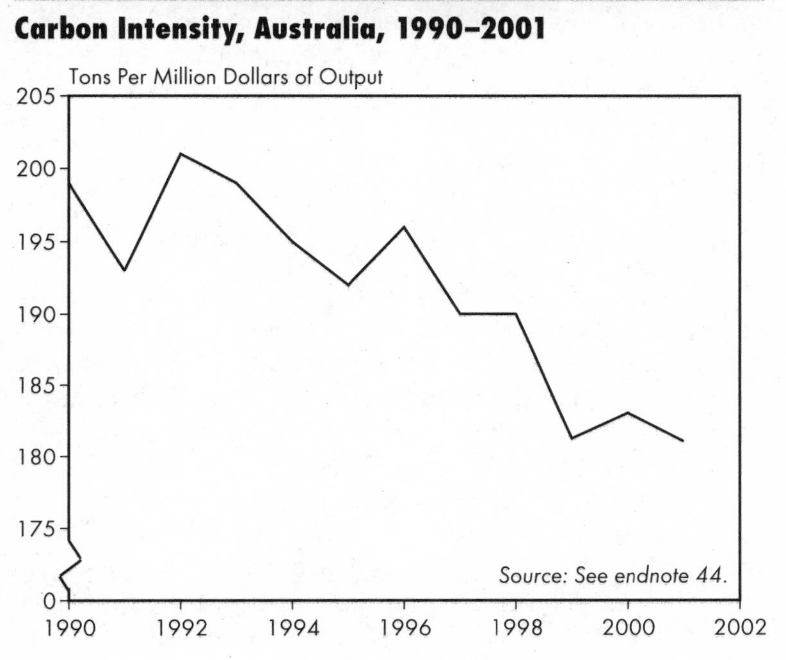

of overall greenhouse gas emissions. Government experts report progress in formalizing agreements with industry and creating a system to monitor their performance. Under the agreements, first signed in 1996, companies prepare inventories and an action plan, forecast emissions levels, undertake monitoring and reporting, verify performance, and make a public statement on the company agreement. Since the program's inception, more than 250 of Australia's largest companies—in energy, resource processing, mining, and manufacturing, and in industry associations—have committed to the GCP, with 116 signatories.[49]

In 1999, these initiatives were supplemented by the Parliament's passage of legislation for a new tax system, with a package of "greenhouse programs" supporting renewable energy and the conversion of vehicles to compressed natural gas. The package includes a Greenhouse Gas Abatement program, the actions of which are being developed with various stakeholders. The government has also established an expert group to explore a possible domestic emissions trading scheme. But many more measures to slow Australian emissions growth will be needed even to meet its generous Kyoto target.[50]

Brazil

Brazil generates a 1.2 percent share of global carbon emissions, and is the sixth-leading carbon emitter among developing nations and 17th among all nations. Its per capita carbon emissions, at 0.5 tons per person, are lower than those of China, less than half the global average, and under one tenth of those of Australia.[51]

Brazilian carbon emissions grew by 58.5 percent from 1990 to 2001. (See Figure 7.) Use of oil, natural gas, and coal increased by 45.7, 288, and 47.4 percent, respectively. Production of oil—which accounts for 49 percent of commercial energy—and of gas also spiked, both by 103 percent. Brazil's carbon intensity rose by 19.4 percent, but remains well below that of China, the United States, Australia, and Canada.[52] (See Figure 8.)

A major reason for Brazil's relatively low carbon intensity is its reliance on modern biomass for 20 percent of its energy and hydroelectricity for 95 percent of its power supply. Maintaining a high proportion of hydropower in electricity generation is thus an important part of the government's climate policy. But the government has also moved to develop a broad range of renewable energy sources, for power generation as well as for transportation.[53]

Since the Rio conference, the Brazilian government has taken steps to promote small-scale hydro, biomass, and wind power, mainly by improving access to power networks, extending financial supports, and reducing bureaucratic demands for registering projects. Recent growth has been rapid: roughly 3,500 megawatts (MW) in renewable energy projects were registered with the Brazilian energy agency in 1998; in 2001

FIGURE 7

Carbon Emissions, Brazil, 1990–2001

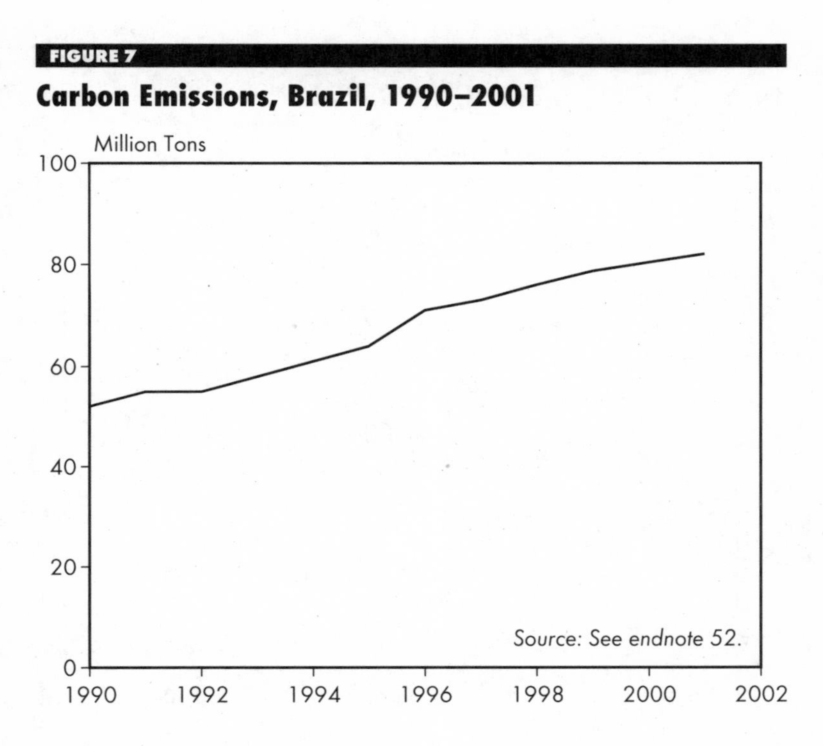

alone, 4,000 MW of wind power projects were registered. While these projects will reduce carbon emissions, their basic premise is "economic rationality," according to the agency. For example, such decentralized power sources can create local job opportunities and provide electrical services in remote areas where they are the only available option.[54]

While not strictly a climate program, an initiative affecting Brazil's carbon emissions has been the "Proalcool" program, based on converting sugarcane to alcohol for use in motor vehicles. Launched in the mid-1970s in response to the first oil crisis, the program mandated the blending of ethanol with all gasoline sold in the country, leading to ethanol use in over 90 percent of new motor vehicles in Brazil by the early 1990s. The program was abating an estimated 12.7 million tons of carbon per year by the mid-1990s, but has been difficult to sustain due to continued low prices and subsidies for oil. By 2000, ethanol's share of motor vehicle fuel consumption had dropped to 20

FIGURE 8

Carbon Intensity, Brazil, 1990–2001

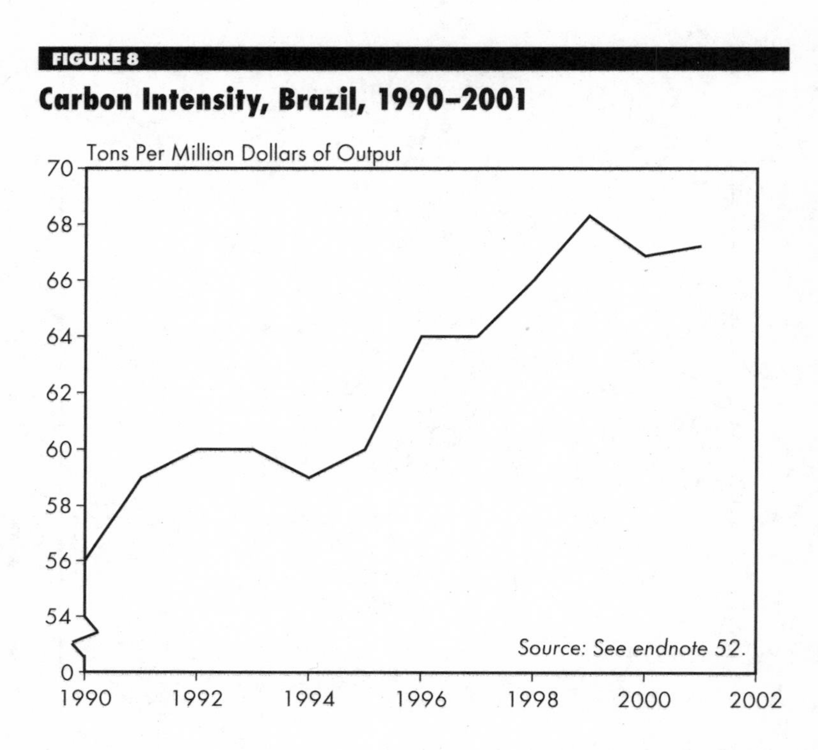

percent—still enough to displace the equivalent of 220,000 barrels of oil per day. Despite its setbacks, the alcohol program has prevented the addition of $140 billion to Brazil's foreign debt, and demonstrates the potential to introduce low-carbon transport fuels.[55]

Recent renewable energy initiatives have been supported by the Global Environment Facility (GEF), which was set up after Rio to finance developing-country projects that provide global environmental benefits. A biomass cogeneration power plant has been demonstrated in the state of Bahia. In São Paulo, a demonstration of nine hydrogen fuel cell buses is in planning. Brazil has also expressed a strong interest in taking part in Clean Development Mechanism projects, likely to focus on energy efficiency and renewable energy.[56]

Brazil's national efficiency program has seen some interesting results. PROCEL, created in 1985 and focusing on electricity efficiency, saw a doubling of energy savings in its first decade. But officials express concern that recent moves to privatize the sector may pay insufficient attention to efficiency measures.[57]

Brazil has two new institutional climate-related innovations. The Interministerial Commission on Global Climate Change was created in 1999 to articulate the government's climate-related efforts, citing the Protocol's potential for mobilizing resources on the order of "many billion dollars a year" toward its national development priorities. In 2000, the Brazilian Forum on Climate Change was established to provide civil society with more information on the subject, through conferences and a website. These are intended to build Brazil's public base of support for stronger climate-related policies.[58]

Canada

Canada is responsible for 2 percent of global carbon emissions, placing it 9th overall. Its per capita emissions, at 4.2 tons per person, are nearly four times the global average and the third highest in the world. Canada committed under the Kyoto Protocol to a 6 percent greenhouse gas emissions reduction between 1990 and 2008–12. But through 2001, its carbon emissions had exceeded 1990 levels by 11.5 percent.[59] (See Figure 9.)

Rising consumption drove the increase, with growth in coal, oil, and natural gas use of 18.4, 13.3, and 17.6 percent, respectively. Other factors include the shutdown of seven nuclear reactors in Ontario and their replacement with coal-fired stations and resumed economic growth, led by energy exports, after an early-1990s recession. While Canadian carbon intensity remains high, it declined by 15.7 percent with the help of strong economic growth.[60] (See Figure 10.)

In 1995, the government produced Canada's National Action Program on Climate Change (NAPCC), which set out strategic directions for stabilizing greenhouse gas emissions at 1990 levels by 2000. It also established the Voluntary Challenge and Registry (VCR), a related voluntary program with indus-

FIGURE 9

Carbon Emissions, Canada, 1990–2001

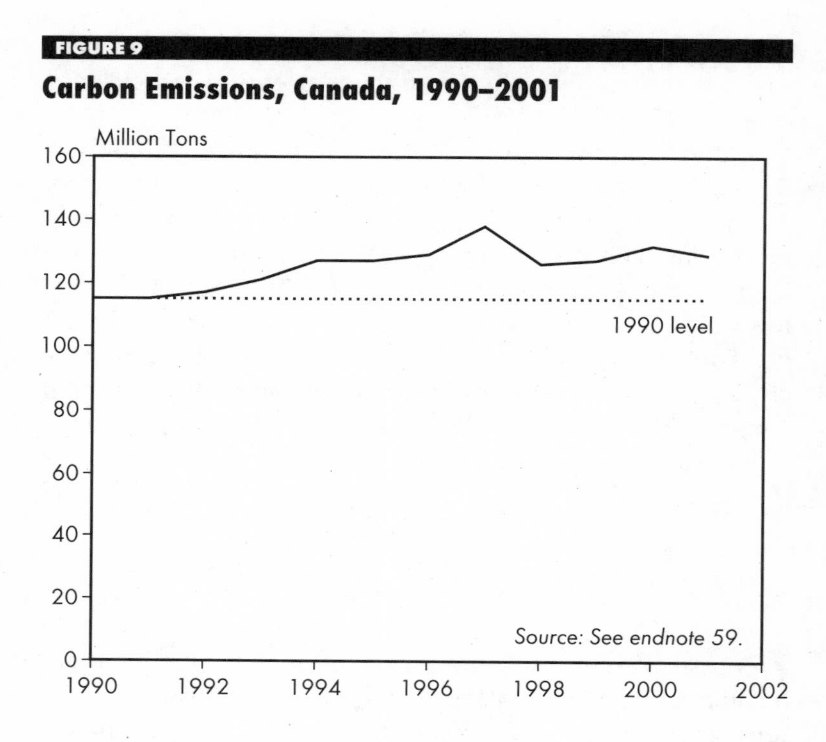

try. While this and other programs were set forth in Canada's 1997 national communication, a 1999 in-depth review of Canadian climate policy found that this communication plus additional information gave "every indication that Canada's GHG emissions will remain substantially above 1990 levels."[61]

Canadian climate policy has evolved since the Kyoto summit, with several policies adopted or strengthened. In December 1997, the Prime Minister and provincial premiers agreed that climate change was an important global issue and that Canada "must do its part." In 1998, the government established a National Climate Change Secretariat to coordinate an implementation strategy; launched a multistakeholder process to examine the options, costs, and benefits of implementing the Protocol; agreed to develop "early actions" to reduce emissions; and signed the Protocol.[62]

The Protocol also led to modest budget increases and several new initiatives. But outside reviewers visiting Canada

FIGURE 10

Carbon Intensity, Canada, 1990–2001

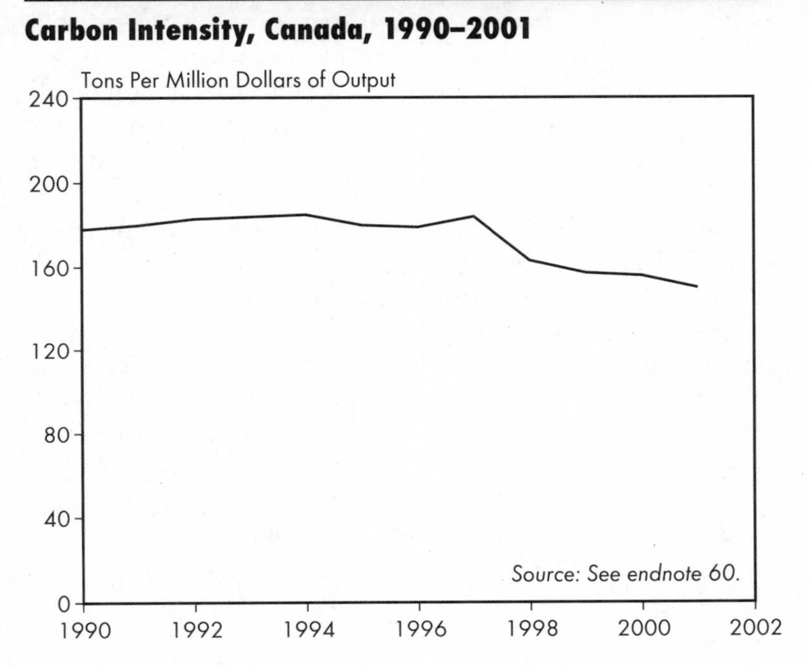

Source: See endnote 60.

found that "mixed signals are being sent by the government on matters related to climate change." While the federal budget for climate-related activities has increased, provincial budgets have largely been cut or held steady.[63]

The outside review team also found that Canada's national climate reports reveal little information on the costs of current or prospective policies. It noted that the actual or potential impacts of such policies were hard to quantify. This is partly because, while regulatory measures, financial incentives, and research and development are among the policies in use, the majority of actions under the NAPCC are voluntary measures—whose effectiveness is inherently difficult to gauge.[64]

Canada's climate plan includes approximately 475 measures, of which the VCR is a main one. The VCR asks (but does not require) companies and governments to develop action plans and submit them to a central registry where they are made publicly available. While more than 700 companies have signed on, the 1999 review team found that there was no adequate system for monitoring and reporting on these initiatives—a problem related to a lack of mandatory reporting.[65]

Since 1995, Canada has spent about $1.27 billion on climate change programs. In 2000, the federal, provincial, and territorial governments agreed to a National Implementation Strategy and a National Climate Change Business Plan, and released the Action Plan 2000 on Climate Change. The last of these will invest $500 million over five years on specific actions to reduce emissions, which when fully implemented are projected to take Canada one third of the way to its Kyoto target. But it is clear that, without additional policies and measures, Canada's carbon emissions will continue to rise beyond its Kyoto target. Addressing this problem will require closer attention to the voluntary programs and more aggressive steps in the transport sector—the area that has the highest growth rate and has received the least attention.[66]

China

China has an 11 percent share of global carbon emissions, the largest among developing nations and the third largest in the world. But its historical contribution to emissions since 1900 is just 7 percent. Its per capita emissions, at 0.5 tons per person, are among the world's lowest, half the global average, and one ninth those of the United States.[67]

Chinese carbon emissions increased by 15 percent from 1990 to 2001, a trend that includes a steady rise followed by a sharp decline and partial recovery. (See Figure 11.) Consumption of oil and natural gas grew during this period by 110 and 88.6 percent, respectively. But coal use dropped by 2.4 percent, with a 23.1 percent decline since 1996. These coal statistics, and consequently the carbon trend, have been questioned by experts suggesting that they exclude the illegal operation of officially "closed" mines. It remains likely, nonetheless, that China experienced some drop in coal use

China's carbon emissions would be 50 percent higher than they are now. Zhang argues that "this makes China's achievement unique in the developing world, and surpasses that of the OECD countries," adding that these achievements need to be better communicated to the outside world to correct the "distorted picture" of China not addressing its carbon emissions or carbon intensity.[72]

Although China's carbon emissions growth may be lower than expected, there are reasons for further reductions, many of which could result in net economic gains. More deliberate climate policies might include further increasing natural gas and renewable energy availability and raising the energy efficiency of electric motors, vehicles, and appliances. China already has over 150,000 solar home systems and 350 MW of wind power in place, and is the leading user of biogas and small-scale hydro and wind power. In addition to lowering emissions growth, these advances also boost productivity and alleviate China's growing local and regional environmental problems such as air pollution and acid rain. China has also expressed a strong interest in the Clean Development Mechanism, which would attract private foreign investment for carbon-saving projects at home.[73]

European Union

The EU and its 15 member states account for 12.7 percent of global carbon emissions, a share roughly half that of the United States, after which it is second, and slightly above that of China. Under the 1997 Kyoto Protocol, which the EU ratified in June 2002, the EU committed to reducing greenhouse gas emissions by 8 percent between 1990 and 2008–12. Through a burden-sharing agreement, member states assumed individual emissions goals.[74]

In April 2002, the European Environment Agency (EEA) reported that the EU had succeeded in reaching its collective Rio target, with a 0.5 percent carbon emissions cut from 1990

FIGURE 13

Carbon Emissions, European Union, 1990–2001

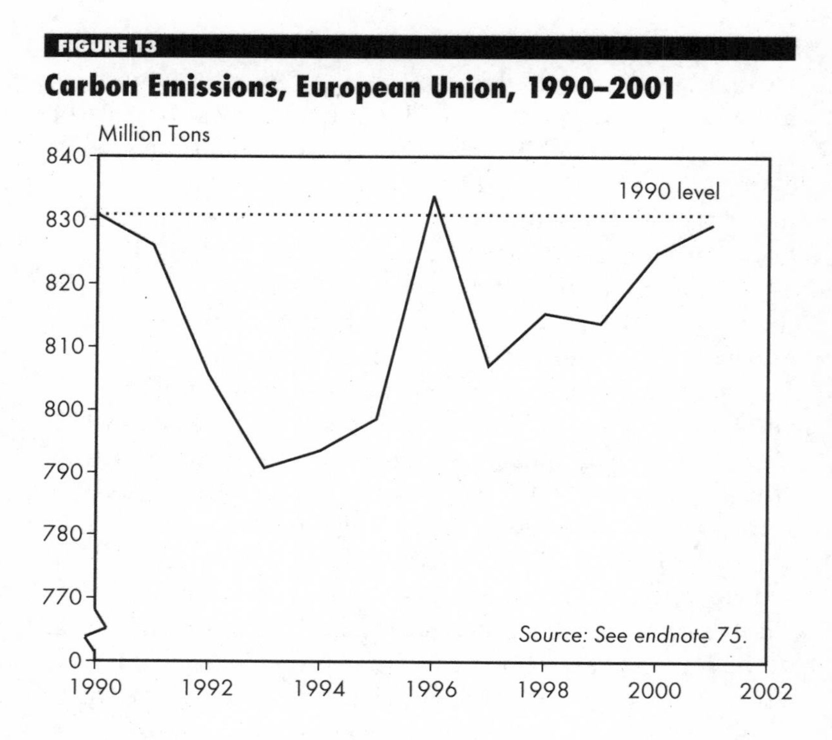

to 2000. Estimates from 1990 to 2001 based on data from the U.S. Oak Ridge National Laboratory (ORNL) and British Petroleum (BP) show a reduction of only 0.2 percent. (See Figure 13.) The EEA estimates that the EU is "halfway to Kyoto," thanks to cutbacks in other gases that resulted in a 3.5 percent greenhouse gas reduction between 1990 and 2000. EU carbon intensity declined by 19.8 percent.[75] (See Figure 14.)

However, carbon emissions, which account for 80 percent of EU emissions, rose in 2000–2001, a worrisome uptick attributed to an increase in coal-fired power generation in the United Kingdom, and continued emissions growth in Greece, Spain, Ireland, Italy, and Belgium. The EEA concluded that the EU faces challenges in resuming its downward emissions trend and meeting its Kyoto goal.[76]

The aggregate EU number masks wide variation in national carbon emissions. Over half of the EU countries are above their target, with the emissions of Spain hitting 34.9 per-

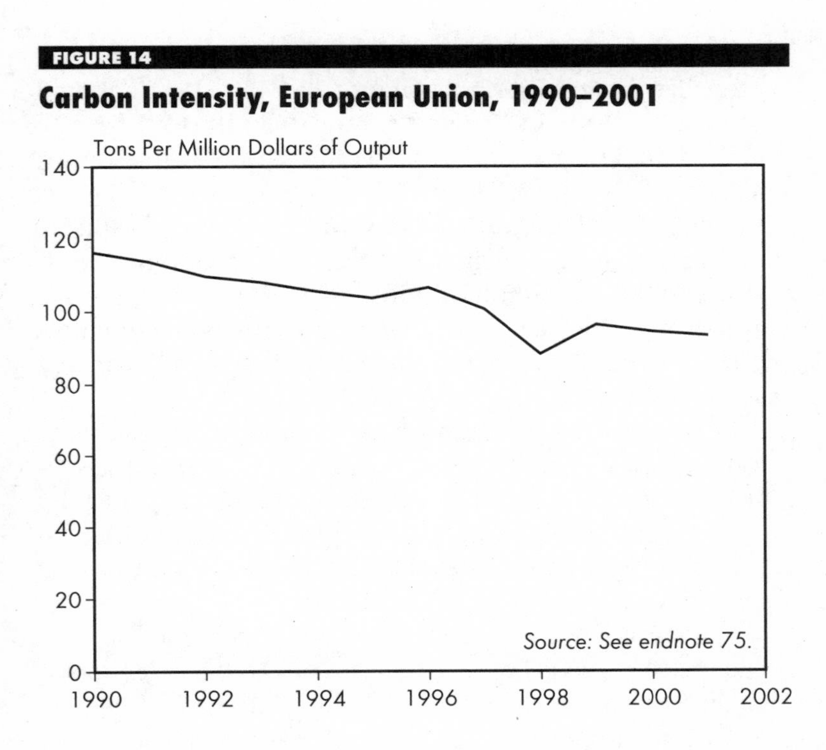

FIGURE 14

Carbon Intensity, European Union, 1990–2001

cent above 1990 levels by 2000, according to the EEA. The best performers have been the EU's leading emitters, Germany and the United Kingdom. Germany recorded a 15.4 percent reduction according to EEA data—and a 17.1 percent cut according to ORNL/BP data. The United Kingdom registered a 7 percent cut under the EEA data, and a 4.1 percent reduction under the ORNL/BP data.[77]

The downward trends in Germany and the United Kingdom can be chalked up mainly to a "dash for gas" from coal to natural gas for electricity generation in the United Kingdom, and to the "wall-fall" reductions resulting from the reunification and closing of inefficient industries in Germany. In the United Kingdom, coal use dropped by 37.1 percent and natural gas use rose by 82 percent between 1990 and 2001. In Germany, coal use fell by 34.9 percent and natural gas use rose by 38.4 percent.[78]

While the European Commission proposed a Community-wide energy tax in 1992 and 1997, an EU-wide levy did not materialize in the 1990s. Several member states have adopted carbon taxes: Denmark, Finland, Ireland, Italy, the Netherlands, and Sweden. Denmark has reported noticeable reductions in energy use from its tax. Though Norway is not in the EU, its 1991 levy has reduced carbon emissions from power plants by 21 percent. Swedish officials report that carbon dioxide emissions in 2000 were 5 million tons less than they would have been without the changes in energy and carbon taxation during the 1990s. The impact of these taxes has been limited by exemptions for industry, however.[79]

The Union's ongoing integration and liberalization of national energy markets led to the near elimination of coal subsidies in Belgium, Portugal, and the United Kingdom and a significant increase in natural gas demand. EU-wide natural gas use grew by 53.3 percent between 1990 and 2001, while coal consumption dropped 27.9 percent.[80]

The promotion of cogeneration and renewable energy is a major element of EU climate policy. In 1999, the EU endorsed a target of doubling the amount of energy production derived from cogeneration, from 9 percent to 18 percent, by 2010. The Commission has also set a strategy and action plan for doubling the contribution of renewable energy to overall energy supply, from 6 to 12 percent, by 2010.[81]

The transportation sector has received less attention in EU climate policy, though transport-related carbon emissions rose from 20 to 25 percent of EU carbon emissions between 1990 and 2000. This has been attributed to a shift in freight from rail to road and to a decrease in the fuel efficiency of cars. The Commission has reached voluntary agreements with European, Japanese, and Korean car manufacturers to achieve a 25 percent reduction, between 1995 and 2008–09, in carbon emissions from cars sold in the EU.[82]

In 2000 the EU established a European Climate Change Programme to identify cost-effective policies, and in 2001 it submitted a set of additional measures needed to implement the Kyoto Protocol, including the promotion of energy efficiency and renewables. It also released a proposal for a directive that would create mandatory greenhouse gas emissions trading within the EU. The program would have a preliminary phase in 2005–07, and a second phase during the Kyoto commitment period of 2008–12. While this may lead to a pan-European program, a major challenge ahead will be to coordinate the scheme with domestic trading programs launched by Denmark and the United Kingdom, which operate under different rules and involve different economic sectors. The EU program is mandatory, excludes the chemical sector, and initially covers only carbon emissions; the Danish program is mandatory, includes only the power sector, and covers only carbon emissions; and the U.K. program is voluntary, excludes the power sector, and covers all greenhouse gases.[83]

"Business-as-usual" projections—which assume no new carbon or energy taxes—show EU carbon emissions rising to 8 percent above 1990 levels by 2010. An in-depth review of EU climate policy, conducted in 2000 by a team of outside experts, observed that in terms of national implementation of EU policy, "progress...has been slow." Noted the International Energy Agency (IEA) in 2001, "It is unlikely...that circumstances which helped the EU return its emissions to 1990 levels during the 1990s will suffice" to achieve Kyoto's 8 percent reduction.[84]

Germany

Germany leads the EU in carbon emissions, accounting for 26 percent of the regional total and 3.3 percent of the global total. Under the Kyoto Protocol, the government committed to a 21 percent emissions cut below 1990 levels by 2008–12. The German government has also established a national target of reducing carbon emissions by 25 percent between 1990 and 2005.[85]

Germany's carbon output dropped 17.1 percent between 1990 and 2001. (See Figure 15.) This decline was related primarily to the 1989 collapse of the Berlin Wall and the reunification of the two Germanys and the subsequent economic

FIGURE 15

Carbon Emissions, Germany, 1990–2001

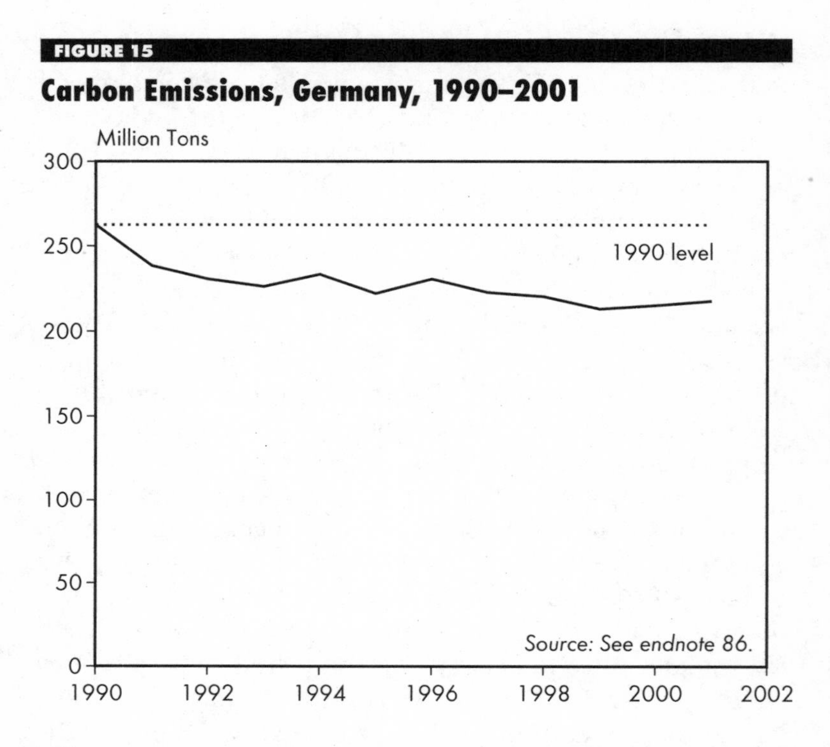

slowdown. The ensuing shutdown of inefficient industries in the former East Germany, a decline in coal consumption, and a shift to natural gas for electricity use in the region put Germany on track to achieving its Kyoto target. However, the large declines in energy consumption in the early 1990s were also related to initiatives for reducing carbon intensity, which declined by 31.9 percent.[86] (See Figure 16.)

Germany's carbon emissions trend encompasses two different patterns. In former West Germany, carbon emissions grew steadily, paralleling increases in energy consumption. In former East Germany, carbon emissions plummeted as the economy was restructured, energy efficiency improvements were made, and coal use was reduced. But while virtually all of the reductions were in former East Germany, they should not be attributed only to the economic downturn. A federal government study suggests that energy efficiency measures in the region's residential sector have also helped.[87]

FIGURE 16

Carbon Intensity, Germany, 1990–2001

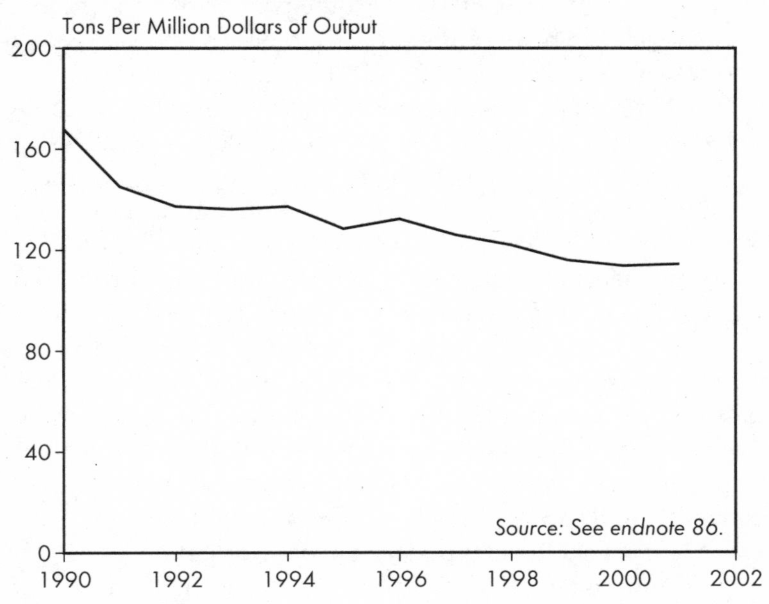

The German government has reported more than 130 policies and measures at the national, regional, and municipal levels. The most common of these are economic instruments, with regulations also playing an important role. However, the federal government has agreed to postpone further regulatory measures, giving priority instead to voluntary commitments from industry—a major piece of Germany climate policy.[88]

German voluntary agreements, or VAs, with industry stem from a 1995 declaration by the Federation of German Industries (BDI) to reduce CO_2 emissions and energy consumption by 20 percent between 1990 and 2005. Specific industry commitments range from a 16 percent cut by the iron and steel association to a 66 percent cut by the potash association. Some targets involved emissions per unit of output, but 12 involve absolute emissions reductions. Altogether the agreements cover two thirds of German industry's energy use, and nearly all of the electricity generation.[89]

Compliance with the industry commitments is monitored by an independent agency, and some promising results have been reported. The manufacturing and electric power sectors had achieved reductions of 27 and 17 percent, respectively, by 2000. Some goals are close to business as usual, and some reductions are unrelated to the agreements. But an outside U.N. review team has lauded the agreements' high level of independent monitoring and objectivity, concluding that "this experience could be usefully replicated by other countries."[90]

European energy market liberalization has prompted Germany to trim coal subsidies, which had propped up domestic prices to nearly three times the world price. In 1997, the government agreed to progressively reduce hard coal subsidies between 1996 and 2005, from $4.2 billion to $2.6 billion. Even more important has been the removal of the Kohlepfennig, or coal levy, paid by West German consumers through their electricity bill. Since this change, subsidies have been paid directly from the federal budget to coal companies.[91]

Promotion of renewable energy is another element of German climate strategy. In 1991, the government passed an act requiring power suppliers to purchase electricity from renewable electricity at a minimum purchase price. The government has also directly subsidized renewable energy, including solar, wind, biomass, and hydropower. Wind power has benefitted most from these supports, growing in capacity from 61 MW in 1990 to over 8,700 MW in 2001, the largest wind power capacity in the world. The government has also launched a "100,000 Solar Roof" program, and introduced minimum prices for selling solar photovoltaic (PV) electricity to the grid.[92]

However, outside experts question whether climate change concerns are being more broadly integrated into German policymaking. They also note that Germany's reporting of climate policy is strong on facts and figures, yet relatively weak on analysis. And there is evidence that Germany will need to take new steps to continue to lay claim to climate policy leadership. Absent new measures, carbon emissions are projected to rise to 1 percent above 1990 levels by 2010; with some new measures, emissions are projected to reach 16 percent below 1990 levels by 2010. In neither case would Germany meet its Kyoto target.[93]

United Kingdom

The U.K. is the EU's second-leading emitter, with 18 percent of the regional total, and accounts for 2.2 percent of global carbon emissions. Like Germany, the U.K. has reached its UN FCCC goal of returning carbon emissions to 1990 levels by 2000. U.K. carbon emissions fell by 4.1 percent during this period. (See Figure 17.) However, emissions rose in 2000 and 2001 due to an increase in coal-based power generation resulting from a temporary ban on new natural gas plant construction.[94]

The government has also established a domestic goal to reduce emissions of CO_2 by 20 percent by 2010. Under the Kyoto Protocol, the country is committed to a 12.5 percent cut in greenhouse gas emissions between 1990 and 2008–12. Cuts in other gases have reduced overall U.K. emissions by 12 percent—nearly to the Kyoto goal. U.K. carbon intensity fell by 23.9 percent.[95] (See Figure 18.)

The most important factor in U.K. carbon trends to date has been privatization and deregulation of energy markets. During the 1990s, natural gas became the fuel of choice for electric power generation, assuming a greater role in industry and the commercial and residential sectors. Use of coal, which had been the primary fuel for electricity, declined as it was replaced by natural gas. As noted above, from 1990 to 2001, this dash for gas resulted in a substantial rise in natural gas consumption and a dramatic drop in coal consumption.[96]

U.K. climate policy also promotes renewable energy and cogeneration. The Non Fossil Fuel Obligation, requiring suppliers to contract for a certain amount of non-fossil power, is funded by a fossil fuel levy and has increased renewable energy use. But the government fell well short of its goal of 1,500 MW of new renewables in 2000, reaching about 700 MW. In 1999

FIGURE 17

Carbon Emissions, United Kingdom, 1990–2001

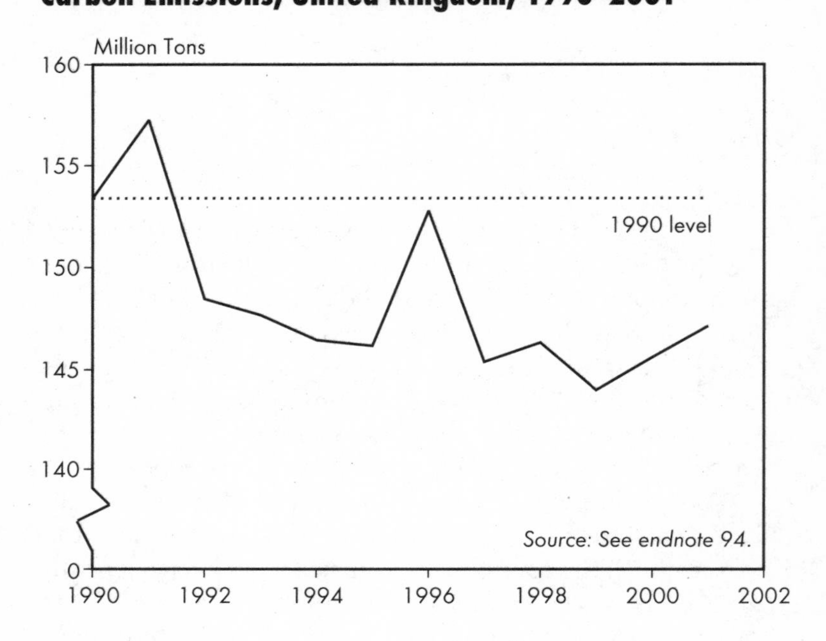

it set new targets, of 5 and 10 percent of total generation by 2003 and 2010. The U.K. came closer to its goal of 5,000 MW of cogeneration by 2000—reaching 4,000 MW.[97]

Another program that fell short was the U.K. Energy Savings Trust, whose budget was slashed and which achieved only one quarter the expected emissions cuts. But overall, most policies have been on track to achieve the expected emissions cuts. Regular increases in the road fuel duty, for example, have reached the 2000 target and are expected to save 1–2.5 million tons of carbon annually by 2010.[98]

Beginning in 1997, U.K. climate policy moved from an emphasis on energy market liberalization and voluntary measures to a broader approach with a greater role for economic instruments. But while progress in switching to cleaner energy supplies has been promising, progress in transport, energy efficiency, and renewable energy has been slow. One positive outcome was a modest levy on business energy use.[99]

FIGURE 18

Carbon Intensity, United Kingdom, 1990–2001

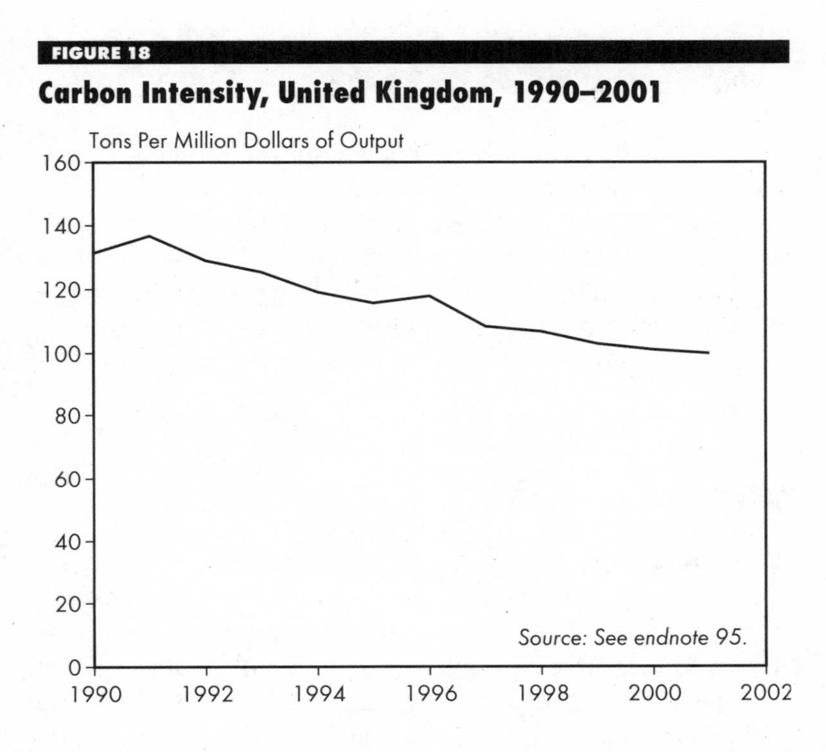

In 2001 the U.K. government introduced a broad policy package, including a climate change levy on business; new energy efficiency standards for appliances and buildings; a renewables obligation of 10 percent by 2010; a goal of doubling the combined use of heat and power by 2010; a 10-year plan to improve public transit; and the world's first broad-based national emissions trading scheme. The last of these policies will provide up to $312 million between 2003 and 2008 to encourage British firms to sign up to emissions reduction targets. The government estimates that the program, launched in April 2002, could be cutting 2 million tons of carbon per year by 2010, while giving industry a global competitive advantage by generating new job and investment opportunities. As noted above, the trading scheme is voluntary, includes all sectors except for electricity, and covers all greenhouse gases. Firms that sign up receive financial incentives or exemptions from the climate change levy. Despite some weaknesses, the overall package has prompted *The Economist* to boast that "Britain leads the world in tackling climate change."[100]

The recent uptick may augur future U.K. carbon trends, however, as switching to gas becomes more difficult, and as energy demand grows. Indeed, a team of expert reviewers visiting in 1999 concluded that beyond 2000 "an upward tendency is anticipated.... new policies appear necessary together with strengthening of some of the existing policies, to achieve the national aim and the binding targets under the Kyoto Protocol."[101]

India

India produces a 4.6 percent share of global carbon emissions, placing it sixth overall and second among developing nations. Its per capita emissions of 0.3 tons per person are the lowest among large emitters, less than one third of the global average, and nearly one nineteenth those of the United States. India's historical contribution to carbon emissions since 1900 stands at 2 percent.[102]

As expected, India's population, energy use, and economy all grew from 1990 to 2001. Indian carbon emissions increased by 71.4 percent. (See Figure 19.) Driving this trend was growth in consumption of coal, oil, and natural gas by 63.4, 67.7, and 111.6 percent, respectively. India is the third-leading coal user, with 7.7 percent of the global total, and coal accounts for 55.1 percent of the nation's commercial energy use.[103]

The government of India undertook a number of climate-related activities during the 1990s. Indian government initiatives since 1990 have succeeded in restraining the nation's carbon intensity, which fell by 5.6 percent. (See Figure 20.) India's carbon intensity is now slightly below that of China. Many of the effective changes relate to the liberalization and fiscal measures adopted by the Indian government in 1991, which have moved the energy sector toward a more market-based pricing system. The initiatives encouraging this trend

FIGURE 19

Carbon Emissions, India, 1990–2001

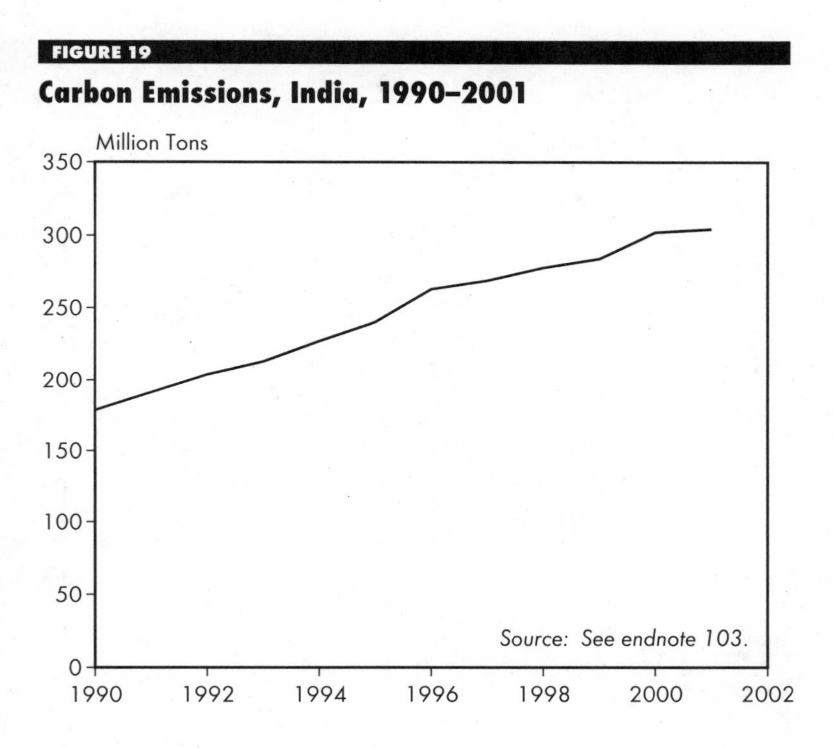

include changes in pricing structures and reductions in energy subsidies, which have improved energy efficiency in the service, industry, and transport sectors. Also important in limiting greenhouse gas emissions has been India's promotion of renewable energy, particularly for rural areas.[104]

India has developed one of the world's largest renewable energy programs. The government established the independent Ministry of Non-Conventional Energy Sources (MNES) in 1990, and in 1992 moved away from state-sponsored programs to the commercialization of renewable energy technologies. The government also has a separate financial institution—the Indian Renewable Energy Development Agency—to promote, develop, and finance these technologies. India now has over 3,400 MW of biomass, solar PV, small hydro, and wind power installed—more than 1,500 of which are from wind, making India the fifth largest wind producer. India also has more than 450,000 solar home systems in place.

FIGURE 20

Carbon Intensity, India, 1990–2001

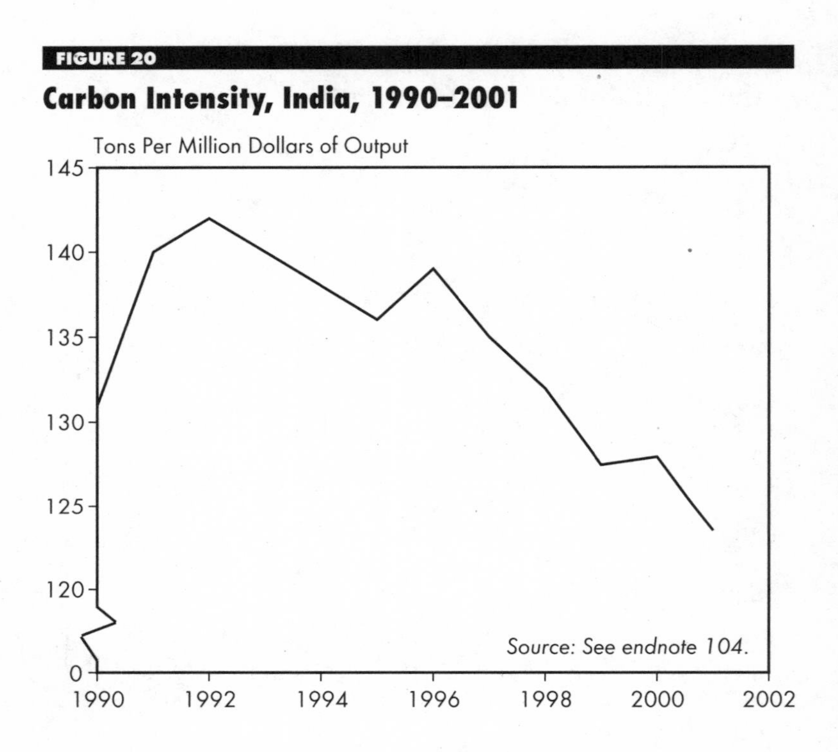

Under the latest MNES strategy and action plan for renewables, a high priority has been given to grid-quality power and rural electrification, and to encouraging private sector involvement through demonstrations, financing by IREDA of viable projects, and fiscal incentives. But there remains significant unrealized potential, due to high initial and transaction costs, underdeveloped markets, and the pricing of conventional energy sources.[105]

Climate change has begun to embed itself in Indian government policy. While global environmental issues were not a priority in the 1992–97 Five Year Plan, the 1997–2002 plan acknowledged their growing importance. Reporting of carbon emissions, energy conservation and efficiency, increased natural gas and renewable energy use, and the phasing out of conventional subsidies became part of stated policy. The government has also introduced natural gas vehicles, GEF projects for fuel cell buses and biomass use, and bilateral efficiency and biomass projects that may serve as precursors to CDM projects.[106]

Rajendra Pachauri, President of the Tata Energy Research Institute and the new chair of the IPCC, notes that energy efficiency and renewable energy are key aspects of the Indian government's development efforts and mandate to alleviate poverty. He observes that in many developing countries, measures like fuel price reform, energy sector deregulation, and the promotion of efficiency and renewables, while not aimed primarily at cutting emissions, "have had significant ancillary benefits with regard to climate change." Pachauri concludes that "though India, like other developing countries, has not taken specific commitments to mitigate CO_2 emissions, it is making progress in this direction," adding that U.S. demands for developing-country commitments willfully ignore these signs of progress.[107]

India's trend toward greater integration of climate and energy policy may well accelerate. In a year 2000 agreement with the United States, India set the goal of a 10 percent share of renewable energy in power additions by 2012 and a 15 percent energy efficiency improvement by 2007–08. In early 2002, the government announced a new electricity law intended to increase clean energy production by 10,000 MW by 2012. Six thousand megawatts are expected from wind power by then, primarily for rural villages comprising most of the 76 million homes lacking access to electricity.[108]

Japan

Japan is the fifth-leading emitter of carbon, with a 4.8 percent share of the global total. As of 2001, Japanese carbon emissions were 10.8 percent above 1990 levels. (See Figure 21.) At the same time, Japan has achieved the lowest carbon intensity of any economy, largely due to its higher level of energy efficiency. Japanese carbon intensity declined by 2.4 percent.[109] (See Figure 22.)

FIGURE 21

Carbon Emissions, Japan, 1990–2001

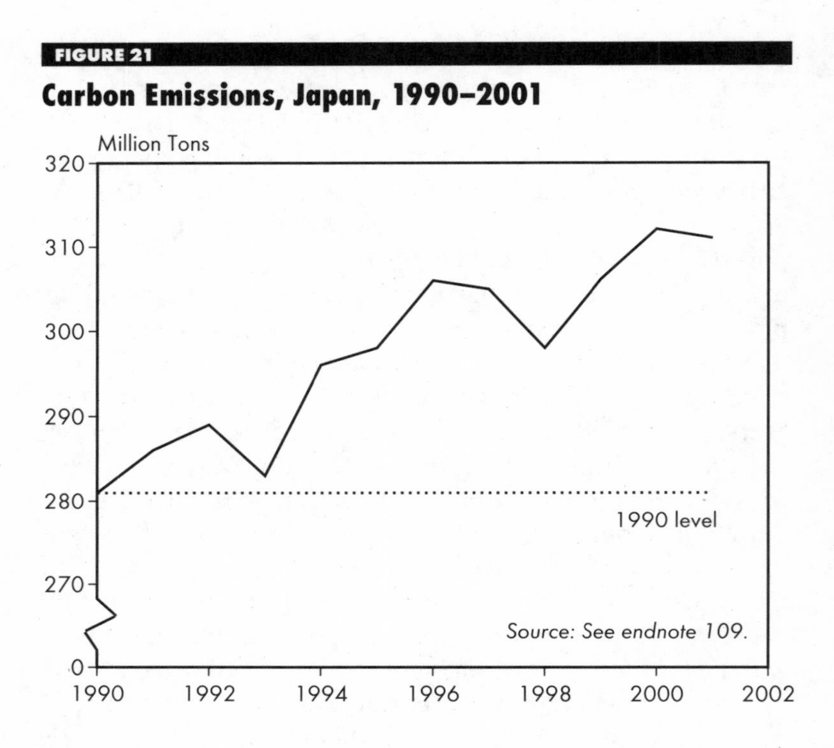

Under the Kyoto Protocol, which Japan ratified in June 2002, the nation is committed to a 6 percent emissions cut between 1990 and 2008–12. But Japanese emissions have risen nearly every year since 1990, an increase driven largely by growth in energy consumption. Consumption of coal and natural gas rose by 35.5 and 54.2 percent, respectively.[110]

The Japanese Action Programme lists more than 400 measures in energy, industry, commercial and residential sectors, and transport. Its overall emphasis is on stringent energy efficiency standards—most of them voluntary—government-supported research into the development of new technology, and government subsidies for the introduction and commercialization of new energy sources.[111]

The most significant climate-related measures within industry are the voluntary action plans of the Japan Federation of Economic Organizations (Keidanren), the nation's largest industry association. The Keidanren projects that these action

FIGURE 22

Carbon Intensity, Japan, 1990–2001

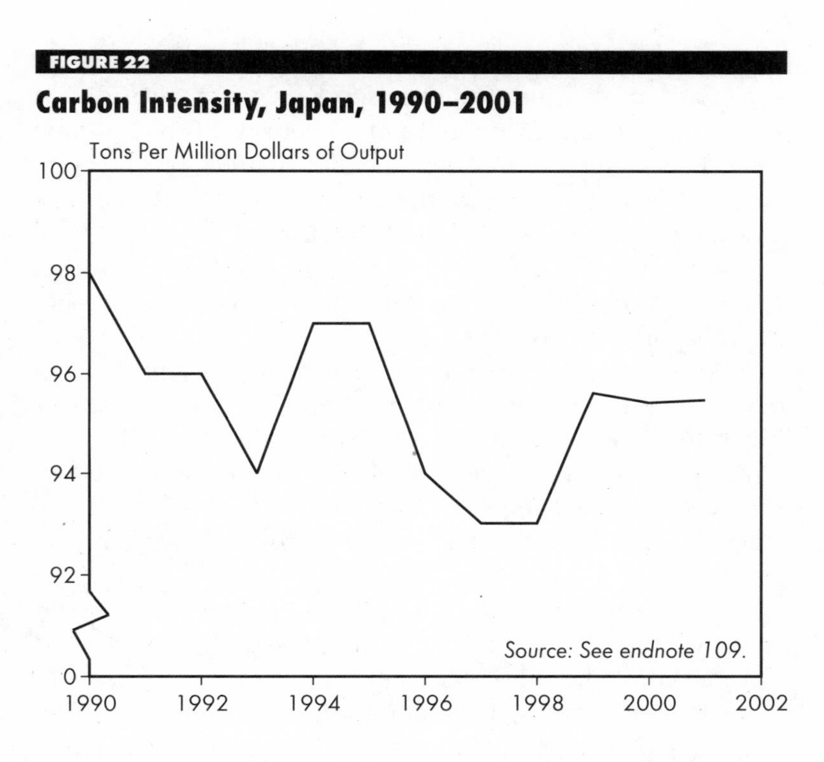

plans will lower industrial carbon emissions to below 1990 levels and save 17 million tons of oil equivalent between 1990 and 2010. A 1998 inhouse review reported that 39 industries and 140 industry groups were participating in the program. Thirty of the industries had set forth quantitative targets for reducing carbon emissions, and associated measures for meeting these targets; and 28 are in energy conversion and industry, which account for more than 40 percent of national carbon emissions. According to the report, 1998 carbon emissions from participating industries were 2.4 percent below 1990 levels.[112]

Since the 1998 review, two new industries have joined the action plan, two industries have revised targets, and two industries have added new 2010 targets. The broader plan is itself being revised to address the Kyoto target, and its progress will be reviewed by the Advisory Council of Ministers. But the review also identified challenges in promoting additional energy conservation steps, since many industries are already highly energy efficient.[113]

Carbon emissions from transport are growing steadily in Japan and account for more than 20 percent of the national total. The Energy-Saving Law, amended in 1998, sets fuel efficiency targets for automobiles that aim for 15–20 percent improvements between 1995 and 2010. The 1998 climate policy also promotes energy-efficient car ownership through reforms in the automobile tax system and improvement of public transport systems.[114]

Renewable energy is another element of Japanese climate policy. Japan's 1998 guidelines set forth goals for increasing wind power capacity from 20 MW to 300 MW by 2010, and solar PV from 81 MW to 500 MW by 2010. These goals are supported through subsidies to renewable energy businesses and through financial and tax incentives for investing in fuel switching.[115]

Japan's solar PV policy is worth noting, as it has driven worldwide development of the solar market. The government introduced a new subsidy system for solar home systems in 1996. Now the leading PV producer, Japan installed 120 MW in 2001. Japan also has more than 30 MW of phosphoric acid fuel cells installed, the highest of any country, and over 500 MW of geothermal plants in operation.[116]

Based on the extrapolation of recent trends, Japan's carbon emissions are projected to reach 20 percent above 1990 levels by 2010. A 1998 energy outlook, however, includes a "policy case" scenario in which reduced oil dependence, continued improvement in energy conservation, and promotion of natural gas and renewable energy return energy-related carbon emissions to 1990 levels by 2010.[117]

Japanese climate policy to date is highly unlikely to achieve the more ambitious Kyoto reduction target. A team of outside experts, reviewing Japan's policy in 2000, noted that the energy and climate policy packages passed in 1999, featuring financial incentives such as subsidies and tax breaks, "considerably strengthen Japan's response to the greenhouse effect." But the team remained "concerned about the need for a more systematic and rigorous approach to monitoring and assessing the effects of these policies and measures, as responsibility for implementing measures is spread across a large number of central and local government agencies."[118]

Russia

Russia is the fourth-leading carbon emitter, with a 6.2 percent share of the global total. Russia agreed under the Kyoto Protocol to maintain emissions at 1990 levels by 2008–12. But its carbon emissions dropped by 30.5 percent between 1990 and 2001.[119] (See Figure 23.)

These trends are linked to the enormous economic and political reforms under way in the nation since 1991—the transition from a centrally planned to a market-driven economy, the major economic downturn through 1997, and the brief recovery and financial crisis. These changes had a profound impact on energy consumption and carbon emission patterns: between 1990 and 2001, Russia's use of coal, oil, and natural gas plummeted by 36.5, 51, and 11.3 percent, respectively. They also pushed Russia's high carbon intensity up slightly, by 0.4 percent.[120] (See Figure 24.)

In 1994, the Russian government set up an Inter-Agency Commission on Climate Change (ICCC) to coordinate national climate policy and represent the country in international negotiations. In 1996, the ICCC launched a Climate Programme, a significant component of Russian climate policy, adopted by government decree and containing a portfolio of research, monitoring, mitigation, and adaptation strategies. However, funding has been reportedly "marginal" in the wake of the 1998 financial crisis.[121]

The foundation of Russian climate policy, as outlined in its 1998 report, focuses on opportunities for energy efficiency improvements and switching to natural gas. It includes a number of energy efficiency and energy strategy programs created since 1995. However, an expert team of outside reviewers reported in 2000 that during their visit "very limited infor-

FIGURE 23

Carbon Emissions, Russia, 1990–2001

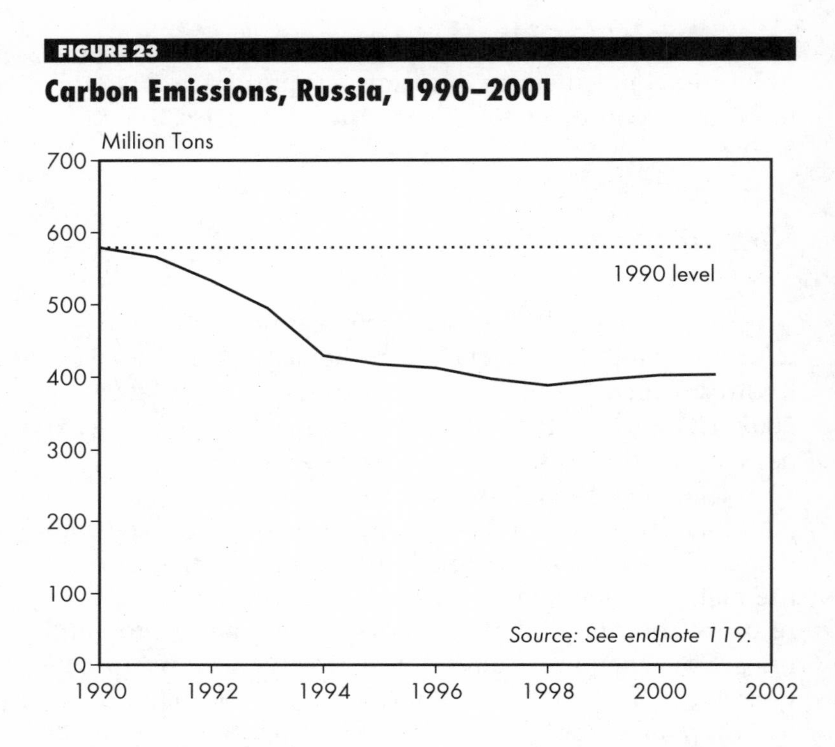

mation was provided on the implementation of these programmes and laws," making it difficult to conduct a thorough analysis.[122]

Another problem with evaluating Russian climate policy, according to the experts, is that its measures are presented in strategic and conceptual terms, rather than in terms of specific initiatives linked to addressing climate change. Though UN FCCC guidelines request it, no information was given on the type of policy instrument, the way the policy interacts with other policies, or the status of its implementation. The team experienced difficulty in analyzing the impact of the policies and measures that had already been adopted, and in making a distinction between the contribution of these policies and the role of the overall economic decline to the downward emissions trend.[123]

Russia's energy strategy, which calls for a change in energy patterns through energy efficiency and an increased

FIGURE 24

Carbon Intensity, Russia, 1990–2001

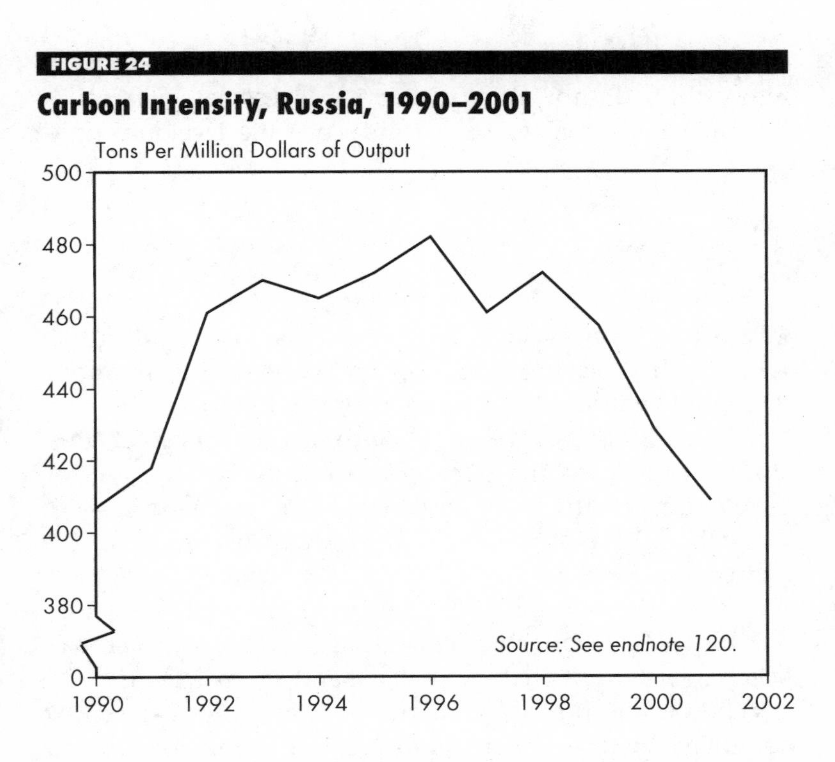

share of natural gas, was revised in the mid-1990s to give as much emphasis to reducing environmental impact as to energy security and efficiency. By 1999, liberalization and privatization of the energy sector had been undertaken, and laws on energy savings and gas supply had been adopted. Under the framework energy-saving law, approximately 30 regional savings laws have been adopted, and numerous energy efficiency funds have been established. Despite the continued subsidization of energy prices, some progress in improving energy efficiency has been made through new building codes.[124]

While new energy efficiency measures have been introduced in recent years, the review team found that "efforts have yet to be made to monitor the implementation and assess the effects, which in turn may strengthen implementation." Without more policies, Russia may hit its Kyoto goal but "have problems with long-term stabilization of emissions" as its economy recovers. And it is unclear if the enormous amount of emissions credits that will accrue to Russia under the Kyoto trading system—dubbed "hot air" by skeptics—may deter efforts to decouple the nation's emissions and economic development.[125]

South Africa

South Africa puts out a 1.4 percent share of world carbon emissions, placing it 15th in the world and fifth among developing nations. Its historical contribution to carbon emissions since 1900 is 1.2 percent. South Africa's per capita emissions, at 2.1 tons per person, are relatively high among developing countries—almost twice the global average—but still below those of the EU, and less than half those of the United States.[126]

Between 1990 and 2001, South African carbon emissions rose by 16.3 percent. (See Figure 25.) Underlying this growth were a 35.5 percent increase in oil use and a 13 percent increase in coal use. South Africa's reliance on coal for 75.3 percent of commercial energy is a major influence on emissions trends. The nation accounts for 3.6 percent of world coal consumption. South Africa also has a 5.6 percent share of coal production, which increased by 36.8 percent.[127]

This legacy of domestic coal dependence is another important factor in South Africa's high carbon intensity, which oscillated somewhat during the 1990s and ultimately declined by just 3.5 percent overall. (See Figure 26.) South Africa's carbon intensity is also relatively high among developing nations—more than three times that of China—and is roughly equal to that of Russia. Other underlying factors include heavy reliance on energy-intensive industries for domestic production and export, higher energy intensity of synthetic petrol made from coal, low energy prices, and poor energy efficiency in most sectors.[128]

Indeed, South Africa has among the lowest energy prices in the world, as well as some of the lowest costs for electricity generation. A 1998 White Paper on Energy Policy pro-

FIGURE 25

Carbon Emissions, South Africa, 1990–2001

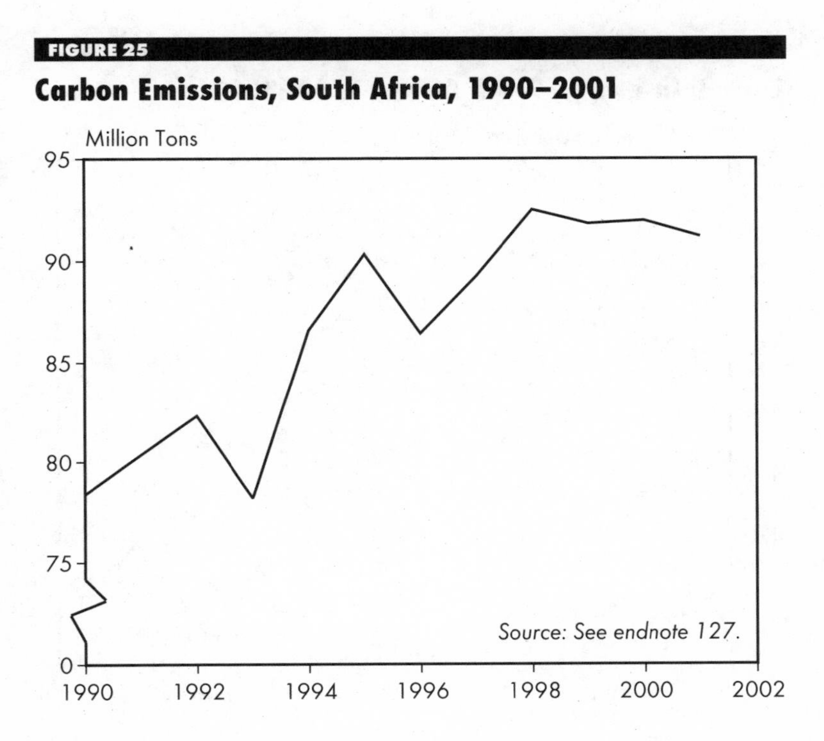

poses several measures that would improve South Africa's carbon intensity. The proposed measures include greater diversity of energy sources, more accurate pricing, the promotion of renewable energy and energy efficiency in a range of sectors, and the development of a natural gas market.[129]

South African public investment in modern, low-carbon energy is still in its early stages. Currently, most renewable energy use is traditional biomass. The South African government's White Paper includes a number of initiatives aimed at promoting "newer" forms of renewable energy. These focus particularly on rural areas, where renewable energy is more cost effective than the extension of the electricity grid from urban areas. Over 50,000 solar units have been installed, and in 1999 former President Nelson Mandela unveiled a concession program aimed at providing 350,000 solar PV systems in remote rural areas. A draft White Paper on Renewable Energy aims to increase the overall share of renewable energy from 9

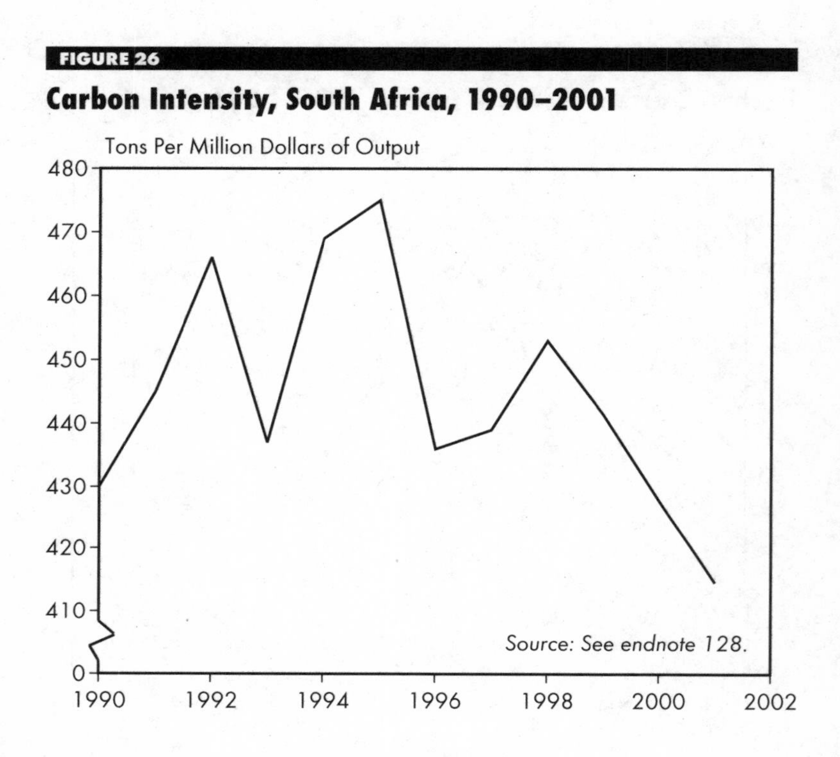

FIGURE 26

Carbon Intensity, South Africa, 1990–2001

to 14 percent between 1999 and 2012.[130]

A looming challenge for South African climate policy will be to reconcile the post-apartheid recognition of equal access to affordable energy with the need for a less carbon-intensive development pattern. As with other developing countries, South Africa's high carbon intensity makes it a potential site for Clean Development Mechanism projects. But new government policies are needed to address the country's carbon intensity in a way that vigorously promotes development.[131]

United States

The United States is the world's leading carbon emitter, with 23.1 percent of the global total—more than double the share of the next national emitter, China, and nearly double that of the EU. It also leads the world in per capita carbon emissions, which, at 5.5 tons per person, are twice those of the EU and Japan, five times the global average, 10 times those of Brazil, and 19 times those of India. It is also the leading emitter on a cumulative basis, responsible for 30 percent of carbon emissions since 1900. As the IEA observes, "Since it is the world's largest emitter of CO_2 emissions, U.S. actions to limit greenhouse gas emissions will be critical to achieving a downward trend in global emissions."[132]

Between 1990 and 2001, U.S. carbon emissions rose by 15.7 percent. (See Figure 27.) During this period, consumption of coal, oil, and natural gas increased by 15.2, 14.6, and 14 percent, respectively. The U.S. shares of world consumption of these fuels now stand at 24.6, 25.5, and 25.6 percent, respectively.[133]

The actual U.S. trend was much higher than projected due to higher-than-expected economic growth and lower-than-expected energy prices. Carbon intensity fell by 17.2 percent, a trend attributed to industrial efficiency and the continued move toward a service, information-based economy. (See Figure 28.) But the U.S. would face great difficulty in meeting its 7 percent reduction target under the Kyoto Protocol, which it has refused to ratify.[134]

U.S. climate policy has been hampered by limited financial support: climate-related funding ranged from roughly 50 to 65 percent of budget requests during the mid- to late 1990s. In addition, the U.S. approach has relied heavily on voluntary instruments, a focus only recently supplemented by proposed R&D increases and financial incentives. And while the government has put in place a system to monitor and report on the progress of its climate action initiatives, an expert team of reviewers visiting the United States in 1998 reported that "the team was not provided with current estimates of the effectiveness of many of the policies and measures."[135]

The U.S. climate action reports show remarkably little change. The first, released in 1993, consisted primarily of voluntary measures, with a small number of regulatory initiatives and no fiscal instruments. The second report, released in 1997,

FIGURE 27

Carbon Emissions, United States, 1990–2001

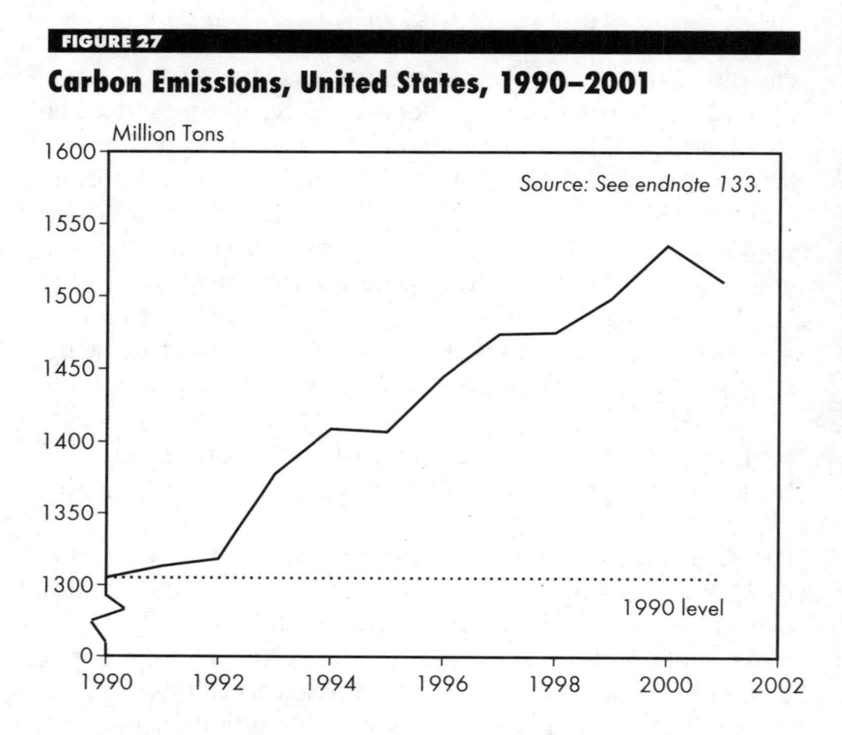

contained only six new policies and measures, which were largely voluntary and were expected to have a minimal effect on emissions. Meanwhile, funding restrictions caused the termination of 11 of the original policies and measures described in the first action report. The third report, released in June 2002, acknowledged U.S. vulnerability to climate change but emphasized adaptation over prevention. It reports a handful of new policies but projects that these would still allow emissions to rise to 34 percent above 2000 levels by 2020.[136]

The report describes and endorses, but does not quantify the effect of, several voluntary programs. The Energy Star labeling program, which promotes energy efficiency in homes and buildings, involves more than 5,500 programs, 500 manufacturers, and more than 200 home builders. The program has been expanded to industry. Climate Challenge involves 650 electric utilities covering more than 70 percent of power generation, but has no targets or outside monitoring.[137]

FIGURE 28

Carbon Intensity, United States, 1990–2001

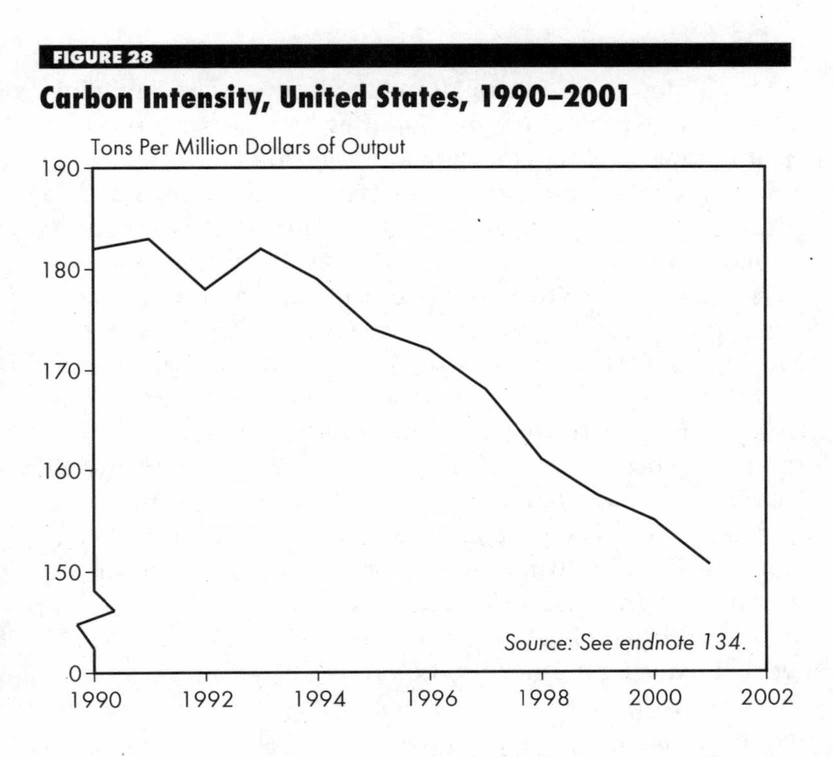

One success has been with appliance standards, which by 2000 had saved 49 million tons of carbon while saving consumers nearly $25 billion. However, stronger appliance standards and building codes have been blocked by Congressional moratoria. Similarly in transportation, efforts to raise Corporate Average Fuel Economy (CAFE) standards—have been stymied by Congress. Outside reviewers have found a "scarcity of measures" in transport that "contrasts starkly" with the sector's contribution to greenhouse gas emissions.[138]

U.S. research and development has led to important cost declines for renewable energy. Commercialization has focused on tax incentives for production and use, which have made the United States the second-leading wind power user. The administration's 2001 energy policy proposal, which favored fossil fuel supports, did create a Clean Energy Initiative, to expand renewable energy markets, work with state and local governments to promote clean energy, and facilitate combined heat and power and distributed power.[139]

Nonfederal U.S. climate policy has been encouraging. One hundred ten cities and counties have drawn up action plans under the Cities for Climate Protection Campaign of the International Council for Local Environmental Initiatives—a campaign reportedly reducing carbon emissions by roughly 2 million tons annually. Massachusetts, New Hampshire, and New Jersey have adopted binding controls on carbon emissions from power plants; California has voted to do the same for motor vehicles. Governors of six New England states have joined with the premiers of five Canadian provinces to commit to returning the region's emissions to 1990 levels by 2010 and to 10 percent below those levels by 2020. Texas and 11 other states have developed renewable portfolio standards, requiring a fixed percentage of electricity sales from renewable sources. And a growing number of U.S. companies have voluntarily adopted emissions reduction targets.[140]

In February 2002, the White House unveiled a national goal of reducing the greenhouse gas intensity of the U.S. economy by 18 percent over the next 10 years, improving the emissions reduction registry under which voluntary reductions are reported, and developing transferable credits for individuals and businesses that register reductions. Under the policy, a review of progress toward the goal would take place in 2012, whereupon new incentives and voluntary measures would be undertaken if necessary. [141]

The plan seeks an increase of $700 million in climate spending, to $4.5 billion, and creates a Cabinet-level Committee on Climate Change Science and Technology. It also aims to implement a range of new and expanded domestic policies, including: tax incentives for renewable energy, cogeneration, and new technology; agreements with industry to improve greenhouse gas intensity and reduce emissions; and transportation programs encompassing tax incentives for consumers purchasing hybrid-electric and fuel cell vehicles, research partnerships aimed at fuel cell vehicles, and the consideration of changes to CAFE standards. Many of these activities have already been suggested in Congress, which has seen a rapid increase in climate-related proposals since 2001, including legislation passed by a Senate committee in June 2002 that would limit power plant carbon emissions.[142]

Critics of the new U.S. plan note that the nation's greenhouse gas intensity fell by 17 percent from 1990 to 2000, yet this failed to restrain rising emissions. They add that a greenhouse gas intensity target will not reduce emissions if not sufficiently stringent: analyses project the goal will result in emissions growth similar to that of the 1990s, an increase of 14 percent by 2010. *The Economist* called President Bush's new goal "utterly inadequate as a target," and his overall climate policy "all hat and no cattle."[143]

Conclusion and Recommendations

Only a few leading emitters—the United Kingdom, Germany, Russia—have met their Rio goals and are on course to meet their Kyoto goals. But as the case studies illustrate, most national governments of industrial countries are stepping up their activity in the area of climate policy. Indeed, the IEA has identified more than 300 separate measures that its members undertook during 1999 to address climate change. The agency placed these actions in five general categories: fiscal policy, market policy, regulatory policy, R&D policy, and policy processes, and noted that "good practice" climate policies should:

- maximize both economic efficiency and environmental protection;
- be politically feasible;
- minimize red tape and overhead; and
- have positive effects on other areas, such as competition, trade, and social welfare.

Using these principles to examine the current record, one can identify several good practices to date.[144] (See Table 3.)

While there is no "silver bullet" climate policy that can be applied across all countries, experience to date suggests that getting the prices right through subsidy reform and tax

TABLE 3

Climate Change Policies and Good Practices, Industrial Nations

Category	Policies	"Good Practices" to Date
Fiscal	Ecotaxes	Norway, Sweden carbon taxes
	Tax credits or exemptions	U.S. wind power tax credit
	Subsidy reform	U.K. coal subsidy removal
Market	Emissions trading	U.K., Denmark emissions trading programs
Regulatory	Mandates/standards	Germany electricity feed law
	Voluntary agreements	Germany, Netherlands agreements
	Labeling	U.S. Energy Star program
R&D	Funding and incentives	Japan renewable energy funding
	Technology development	Japan energy efficiency program
Policy Processes	Strategic planning	U.K. climate change program

Source: See endnote 144.

policy is crucial. Market approaches and a mix of policies—voluntary agreements, standards, incentives, R&D—are also needed. Important as well are monitoring and assessment, good institutions, and international cooperation. Even if their rationale is strong, however, climate policies run into the formidable barriers of perceived high cost and limited political will to act—as has been demonstrated in many of the studies above.[145]

Climate-related fiscal policies have become increasingly popular, with nearly all industrial countries adopting such measures, though most are modest in size. These measures are appealing because they tend to reduce greenhouse gas emissions while stimulating national economies. A good example is the phasedown of coal subsidies in Belgium, Japan, Portugal, and the United Kingdom from more than $13 billion in 1990 to less than $7 billion by the end of the decade. Subsidies are also being added to promote more-efficient vehicles and renewable energy in power generation, the most successful example to date being the German electricity feed law—which has spurred the wind power business and been replicated in several other European nations.[146]

Nineteen industrial nations are planning more than 60 tax policy changes that will affect emissions, although only 11 of these are defined as carbon or emissions taxes. The most effective carbon taxes to date are in Scandinavia: One example discussed in the previous section, Norway's levy, was adopted in 1991 and has reportedly lowered carbon emissions from power plants by 21 percent. One reason such taxes have been adopted slowly or contain exemptions is that their impact on fairness and competitiveness is often overstated by industry.[147]

Interest in market-based mechanisms has also risen, due to their expected cost-effectiveness and the success of the U.S. sulfur emissions trading program, which has helped to reduce sulfur emissions by 24 percent since the program was instituted in 1990. Several countries, along with the European Union, have adopted greenhouse gas emissions trading proposals, and a growing number are considering their adoption. As a recent Pew Center study on Global Climate Change reports, emissions trading is becoming a "policy of choice" for addressing the issue. An international greenhouse gas market is emerging—an estimated 85–105 million tons of CO_2 equivalent have been traded since 1996, nearly half of this in 2001 alone—but in the absence of an international agreement, it is evolving in fragmented fashion. As noted earlier, the U.K., Danish, and EU systems vary considerably in approach, and it will be necessary to reconcile them into a global framework if the trading is to be as economically and environmentally effective as possible.[148]

The third discernible area of growing activity is voluntary agreements, which arise from negotiations between government and business or industry associations. These are attractive because they arouse less political resistance from industry than coercive measures, require little overhead, and can be complemented by fiscal and regulatory measures. Some 21 voluntary agreements were initiated in 1999 by industrial nations, including four for power generation, two for transport, and 11 for industry and manufacturing. With respect to stringency, they are characterized by the IEA as "strong" (in the Netherlands), containing legally binding objectives and the threat of regulation for noncompliance; "weak" (in Canada), lacking penalties for noncompliance but having incentives for achieving the targets; or "cooperative" (in U.S. manufacturing), with incentives for developing and implementing new technology but lacking specific targets.[149]

While voluntary agreements are relatively new, some interesting results have already emerged. As shown above, the German and Japanese business communities have made substantial progress in meeting efficiency targets. Also worth mentioning are the Netherlands' long-term agreements with energy-intensive industry, which achieved a 20 percent energy efficiency improvement between 1989 and 2000. UNEP and the World Energy Council (WEC) have identified more than 700 voluntary projects to cut back greenhouse gas emissions that are just completed, under way, or planned by industry. These have achieved a reduction of 1.3 billion tons of CO_2 equivalent, and a 2 billion ton target has been set for 2010. But UNEP and WEC believe that even as industry activity grows, governments remain too reactive—suggesting that industry could, given the right framework, move much faster than is encouraged by current government arrangements.[150]

While these studies suggest growing engagement by industrial-nation governments in dealing with climate change, the IEA observes that "there remains considerable scope for further improvements." The IEA concludes that policies already enacted and proposed may not suffice for countries to meet their Kyoto targets, and that further action may be necessary. These conclusions are consistent with those drawn from the country studies described above.[151]

In addition to identifying good practices, it is also critical, in evaluating climate policy over the past decade, to compare governments' broader approaches with those recommended by the IPCC literature and described in the second section of the paper. In other words, we need to ask: Did countries take a "portfolio" approach to climate policy, emphasizing a mix of instruments? Did they integrate policies with the non-climate objectives of other social and economic policies? Did they account for the ancillary benefits of policies that cut emissions? Did they coordinate actions internationally? And did they follow the principle that earlier action provides greater flexibility in moving toward long-term goals?[152]

Applying these yardsticks to the country studies, we find that:

- Efforts to develop a balanced, "diversified portfolio" of policies are incomplete, with many governments relying mainly on one type of measure.
- Integration of climate change with the non-climate objectives of other policies has been highly limited.
- There has been little effort to assess the ancillary benefits of policies that reduce emissions.
- Few actions have been coordinated internationally.
- An emphasis on the importance of early action is not evident.

Why have countries largely failed to follow these IPCC recommendations? There are, of course, many reasons that partly explain this divergence between the theory and practice of climate policy. As the previous sections show, climate policymaking is still immature, and the varying quality and quantity of information provided by governments—on the policies developed, their level of implementation, and their actual and projected impacts on emissions—make it difficult to assess what is being attempted, much less what is really being done, and what the impact has been. Therefore, continued progress in the reporting and review of national climate policy is needed. In particular, a more rigorous accounting of the specific emissions impacts of individual policies is required to assess which policies are most effective—and to enable the broader replication of the good practices that do exist.[153]

To the extent that one can accurately compare the climate policies of different countries, it is evident that some are measuring up better than others. The degree of commitment to the climate change issue ranges widely among governments, for several reasons. In some countries, public awareness of the seriousness of the issue is strong, while in others there is still only a vague understanding. In some countries, the issue has broad political support; in others, it divides sharply along party lines. These cultural factors help explain why, for example, the United Kingdom, German, and Japanese policies are clearly more integrated than those of Australia, Canada, and the United States. Equally evident, however, is that all countries could be doing much better with respect to each of the climate policy benchmarks.[154]

We have also learned that effective climate policymaking can be weakened by the misuse of projections. In the case of the United States, overly optimistic projections of business-as-usual emissions trends—based on unrealistic assumptions about economic growth and energy prices—led to the development of inadequate policies. Of these policies, a number were partially implemented or discontinued, resulting in even lower emissions cuts than expected. This combination of events caused U.S. emissions to balloon over the past decade. In the future, governments must resist the temptation to misrepresent the future in order to justify present climate policy.[155]

We have seen as well that climate-related policies can be simply overwhelmed by often longstanding "perverse practices" in the energy and transport sectors. Fossil fuel subsidies in many countries keep the cost of these fuels artificially low and continue to greatly inhibit more climate-friendly patterns of energy use. It is hardly surprising that the countries struggling most with emissions—the United States, Canada, and Australia—exemplify the frontier mentality vis-à-vis climate policy, continuing to distort prices for fossil fuels as if they and the atmosphere's absorptive capacity for carbon were limitless. By contrast, even limited efforts toward subsidy removal and carbon taxation in Europe have yielded results in reducing emissions.[156]

Also falling into the category of perverse practices are the numerous subsidies for road building, suburban development, and car travel that permeate the developed world—particularly

North America, but to an increasing extent Europe as well. The OECD estimates that removing direct and indirect transport subsidies would reduce sectoral emissions by 10–15 percent. Indeed, the transport sector has been a major blind spot in climate policy since Rio, receiving very little attention while becoming the fastest-growing source of emissions. This is a political problem, owing to the breadth of issues bearing on transport and industrial resistance to a strengthening of automobile fuel economy standards. But it will become increasingly important, as transport is projected to remain the fastest-growing source of emissions through 2020.[157]

All countries could be doing much better with respect to each of the climate policy benchmarks.

These perverse practices also create a perceptual problem: Those countries formulating new climate policies have by and large failed to acknowledge those policies that undermine efforts to address climate change, much less advocate their reform. Indeed, in just one country report examined in preparing this paper—that of Sweden—did we find a section discussing "policies that run counter to the objective of reducing greenhouse gas emissions." For climate policymakers to overcome these obstacles, they must do a better job of recognizing their existence more explicitly, and employ economic arguments to overcome political inertia.[158]

Tackling transport, something that industrial nations have yet to do, will be an urgent necessity for developing-nation governments as their climate policies evolve. Transport emissions are projected to grow fastest in the developing world, as these nations continue to experience rapid population growth, urbanization, and increased motorization. Yet there are a range of policies and strategies—road pricing, public transit investment, land use planning—to slow these rates of growth, many of which will also alleviate local air pollution, congestion, and road infrastructure expenditures. This will require learning from the experiences—and mistakes—of the industrialized world in their transportation investments.[159]

Another obstacle to better climate policy has been the reluctance in some quarters to acknowledge the climate-related efforts of developing nations. One of the enduring myths of Kyoto, perpetuated largely by opponents of the Protocol in the United States, is that developing nations would be exempt from any commitments because they lacked the same binding targets. What the case studies suggest, on the contrary, is that, even before such targets are set for them, developing nations are moving to address their emissions—more, some have argued, than many industrial nations. In a 1999 report for the U.N. Development Programme, José Goldemberg and Walter Reid assert that "clearly, developing countries are not passive spectators in the arena of climate change. They have already taken significant steps to reduce their emissions of greenhouse gases below the levels that would otherwise occur." These countries' experiences demonstrate that many steps to reduce emissions make sense on economic grounds alone—a lesson that could be usefully exported from South to North.[160]

But among the many impediments to effective climate policymaking in industrial and developing nations alike, the one looming largest has been the absence of leadership among industrial nations to agreed to binding, specific commitments to reduce emissions. Indeed, the evidence here makes it abundantly clear that the purely voluntary approach of the Rio treaty failed to promote strong domestic climate policymaking over the past decade, and was therefore not up to the job of promoting meaningful progress toward reducing global carbon emissions. This conclusion, in turn, points to the vital importance of a global—and binding—framework to coordinate and accelerate action on climate change.[161]

Increasingly, the need to globally coordinate climate policy has been accepted and advocated by distinguished economists. Joseph Stiglitz of Columbia University argues in a 2001 paper—released the day after he won the Nobel Prize in economics—that significant movement on climate change requires that governments move on two fronts: to adopt cost-effective domestic climate policies and to set an "agenda for global collective action." William Nordhaus of Yale University, who has criticized the Kyoto Protocol as being potentially expensive, nonetheless concedes in a November 2001 issue of *Science* that the treaty's mechanisms will "provide valuable insights on how complicated international environmental programs will work.... It is hard to see why the United States should not join with other countries in paying for this knowledge."[162]

Reengaging the world's largest emitter in the Kyoto process will be difficult, but essential. Richard Schmalensee, Dean of MIT's Sloan School of Management, writes that "the longer the United States, other industrialized nations, and the developing world head down different policy tracks on global warming, the harder it will be to achieve the coordination necessary for effective action." But the purely voluntary approach of the Bush administration seems unlikely to change in the near future, notwithstanding the fact that such an approach, which was already questionable under the first President Bush during the 1992 Rio talks, is far less defensible today, with a decade of policy history under our belts. Indeed, we can now confidently discard the claim made in 1992—and recycled today—that soft, voluntary aims would get us where we need to go. To continue to make this case betrays either policy amnesia or willful neglect of the record of the past 10 years.[163]

U.S. climate policy also places political expediency over economics in ignoring the recent success of its sulfur emissions trading experience. This program set a goal of reducing sulfur dioxide (SO_2) emissions by 10 million tons below the 1980 level, tightening restrictions in two phases. In the first year of compliance, 1995, the program cut sulfur emissions 40 percent below the level required by law. Since then, SO_2 emissions have been dropping steadily, as has the cost of the reductions. To date, the program's cost has been one fifth to one tenth of the $15 billion estimate made in 1990 by the Congress and Environmental Protection Agency.[164]

In its 2002 Economic Report of the President, the administration's own Council of Economic Advisors (CEA) notes that the sulfur emissions trading program "has lowered emissions substantially while yielding considerable cost savings, especially compared with the previous, command-and-control regime." It adds that "as low-cost options for emission reduction emerged that had not been foreseen in 1989, there has been over time a clear downward trend in the predicted cost of the program." Yet the report is surprisingly skeptical about applying these lessons to climate policy, calling an international greenhouse gas trading system "impractical" and arguing that "a flexible international program would be unprecedented." Ironically, CEA Chair R. Glenn Hubbard had proposed, unsuccessfully, the inclusion of mandatory emissions caps and tradeable permits in the administration's climate policy. At the moment, however, Hubbard is limited to abstractly advocating the use of "flexible institutions" to deal with climate change, even as such institutions are being built overseas.[165]

Conventional arguments against the Protocol—that it would be too costly, and that it excludes developing nations—are also belied by our experience in addressing another global environmental problem. It was under the Reagan administration that the U.S. government signed and ratified the 1985 Vienna Convention and 1987 Montreal Protocol to address ozone-layer depletion. As Edward Parson of Harvard's Kennedy School of Government points out, the first round of the Montreal Protocol did not include binding commitments from China, India, and other developing nations. These commitments were phased in during subsequent amendments, and since 1987 the Protocol has achieved a 90 percent reduction in the use of ozone-depleting chlorofluorocarbons, and at a modest cost.[166]

Finally, the common assumption that U.S. businesses will benefit from their government's unwillingness to ratify the Protocol deserves closer examination. In the near term, there may be some advantages over foreign competitors subject to constraints. But over the long term, the ongoing policy uncertainty may have an adverse impact, particularly if other countries' climate policies spur technological innovation, open up new markets, and create a global trading system in which U.S. firms are unable to participate.[167]

There are ways in which separate U.S. and international

strategies might eventually converge. Various proposals for a U.S. national cap-and-trade program for carbon are being considered within the administration and Congress. One of these proposals, from Richard Morgenstern of Resources for the Future, would combine elements of a tax and trading system to allay concerns that carbon prices may skyrocket. Meanwhile, legal amendments to the Protocol could allow the permits that result from such a program to be recognized in the Kyoto system, and for Kyoto permits to be recognized in the U.S. system. But this could create significant complications for multinationals operating within and outside the United States—and could require agreement to certain terms dictated by governments that are already party to the Protocol, just as countries seeking to join the World Trade Organization must demonstrate adherence to certain internationally accepted norms.[168]

The Kyoto Protocol remains the best way to achieve global action on climate change.

These challenges, and the failure of the United States to provide a credible alternative to Kyoto, lend weight to the argument of U.K. climate policy expert Michael Grubb that the Kyoto Protocol remains the best way to achieve global action on climate change. Grubb argues that if the EU leads an international effort, joining with Japan and Russia, to bring the Protocol into force, then the United States will be under greater pressure to rejoin. This would also, he concludes, provide a long-term structure for controlling emissions and strengthen the international framework for continuing action. Further, it would also demonstrate industrial-country leadership, making it easier to bring other nations on board at a later date; and it would bring to the private sector the certainty it seeks—and needs—regarding regulations and targets in order to foster the technological development and spread of energy-efficient and low-carbon technologies.[169]

Indeed, achieving the entry into force of the Kyoto Protocol at the earliest possible date will maximize the options for governments and businesses to map out strategies for meeting

BOX 3

Global Climate Policy Priorities, Johannesburg and Beyond

–Bring the Kyoto Protocol into force.

–Fully account for climate change in reviewing Agenda 21 implementation in the areas of atmosphere, energy, finance, industry, and technology.

–Reaffirm the importance of the IPCC Third Assessment Report as the authoritative starting point for policymakers seeking to implement the Kyoto Protocol.

–Set forth a blueprint for post-Johannesburg climate negotiations, emphasizing the need to reengage the United States, begin discussing the second period of emissions cuts, and expand the group of countries with emissions targets.

–Work to establish a voluntary "Global Climate Compact," modeled after the Global Compact established in 2000 between the U.N. and the private sector, that challenges business leaders to commit to the accelerated deployment of energy-efficient products, renewable energy, and hydrogen and fuel cell technologies.

Source: See endnote 170.

the Kyoto goals and for making progress toward the broader goal of climate stabilization. Bringing the Kyoto Protocol into force is one of several pressing priorities for advancing the global climate change agenda at and beyond the Johannesburg Summit. (See Box 3.) It is also, at this critical juncture, the single most important action needed to strengthen climate policy at both the national and international levels.[170]

International climate policy would also benefit from a specific long-term goal, on which scientists have yet to agree. In a June 2002 issue of *Science*, Brian O'Neill of Brown University and Michael Oppenheimer of Princeton University propose a stabilization target of 450 ppmv—but note that this option would be foreclosed by further delay in reducing industrial-nation emissions. They thus conclude that the Kyoto accord "provides a first step that may be necessary for avoiding dangerous climate interference."[171]

Though climate policymaking is still young, a decade of hindsight has made at least two things clear: Climate change has established itself on the radarscreen of policymakers around the globe, and it will not be going away any time soon. As Thomas Schelling observes in the May/June 2002 issue of *Foreign Affairs*,

> The greenhouse gas issue will persist through the entire century and beyond. Even though the developed nations have not succeeded in finding a collaborative way to approach the issue, it is still early. We have been at it for only a decade. But time should not be wasted getting started. Global climate change may become what nuclear arms control was for the past half century. It took more than a decade to develop a concept of arms control. It is not surprising that it is taking that long to find a way to come to consensus on an approach to the greenhouse problem.[172]

Consensus—at least for this stage of the debate, and for most of the world's governments—may be closer than Schelling thinks. We are living in a moment when the need for multilateral action to address emerging global threats is widely accepted by the international community. Unfortunately, it required the tragic terrorist attacks of September 11, 2001, to shock the public and policymakers out of complacency, and to spur the necessary and long overdue changes in counterterrorism policy at home and abroad. We need not wait for a disastrous climate surprise—a deadly heat wave, a particularly destructive storm, a nearly unmanageable tropical disease outbreak—to move us beyond our current state of complacency and toward the many needed reforms. By implementing the Kyoto Protocol, and by working to further raise public awareness of our vulnerability, we can put the world on a more certain path toward climate stability, and set in motion a second decade of climate policy that builds on the lessons of the first yet is far more successful.

Notes

1. "What is Johannesburg Summit 2002?" at <http://www.johannes burgsummit.org>, viewed 3 May 2002.

2. Text of *UN Framework Convention on Climate Change* (UN FCCC) (United Nations: 1992), at <http://www.unfccc.int>, viewed 3 May 2002; Text of *Kyoto Protocol to the UN Framework Convention on Climate Change* (United Nations: 1997), at <http://www.unfccc.int>, viewed 3 May 2002.

3. Text and Box 1 based on ibid., on Michael Grubb, Christian Vrolijk, and Duncan Brack, *The Kyoto Protocol: A Guide and Assessment* (London: Royal Institute of International Affairs/Earthscan, 1999), on UN FCCC Secretariat, "Governments Adopt Bonn Agreement on Kyoto Protocol Rules," Press Release, Bonn, Germany, 23 July 2001, on UN FCCC Secretariat, "Bonn Decisions to Speed Action on Climate Change," Press Release, Bonn, Germany, 27 July 2001, on UN FCCC Secretariat, "Governments Ready to Ratify Kyoto Protocol," Press Release, Marrakech, Morocco, 10 November 2001, and on UN FCCC Secretariat, "Climate Talks Resume on Eve of Johannesburg Summit," Press Advisory, Bonn, Germany, 3 June 2002.

4. Text and Box 2 based on Colum Lynch, "EU Ratifies Global Warming Treaty," *Washington Post*, 1 June 2001, on Howard W. French, "Japan Ratifies Global Warming Pact, and Urges U.S. Backing," 5 June 2002, on Text of *Kyoto Protocol*, op. cit. note 2, and on UN FCCC Secretariat, "Kyoto Protocol: Status of Ratification," 4 June 2002, at <www.unfccc.int>, viewed 4 June 2002.

5. Quote from "Summary of the Resumed Sixth Session of the Conference of the Parties to the UN Framework Convention to Climate Change: 16–27 July 2001," *Earth Negotiations Bulletin*, 30 July 2001.

6. J.T. Houghton et al., eds., *Climate Change 2001: The Scientific Basis*, Contribution of Working Group I to the Third Assessment Report of the Intergovernmental Panel on Climate Change (Cambridge, U.K.: Cambridge University Press, 2001).

7. Ibid.

8. Figure 1 based on J. Hansen et al., Goddard Institute of Space Studies, "Global Temperature Anomalies in .01 C," at <http://www.giss.nasa.gov/date/update/gistemp>, viewed 25 January 2002; Houghton et al., op. cit. note 6.

9. Houghton et al., op. cit. note 6, p. 10.

10. Ibid., pp. 2–4.

11. Figure 2 based on C.D. Keeling and T.P. Whorf, "Atmospheric CO_2 concentrations—Mauna Loa Observatory, Hawaii, 1958–2000 (revised August

2001)," Scripps Institution of Oceanography, La Jolla, CA, 13 August 2001; Houghton et al., op. cit. note 6, pp. 5–10.

12. Houghton et al., op. cit. note 6, pp. 5–10.

13. Ibid.

14. Ibid., pp. 12–16.

15. Ibid., p. 15; James J. McCarthy et al., eds., *Climate Change 2001: Impacts, Adaptation, and Vulnerability*, Contribution of Working Group II to the Intergovernmental Panel on Climate Change (Cambridge, U.K.: Cambridge University Press, 2001), pp. 3–6.

16. Ibid., pp. 5–7.

17. Ibid., p. 6.

18. Ibid., pp. 6–8; Robert T. Watson et al., eds., *The Regional Impacts of Climate Change: An Assessment of Vulnerability*, A Special Report of IPCC Working Group II (Cambridge, U.K.: Cambridge University Press, 1999); National Assessment Synthesis Team, *Climate Change Impacts on the United States; The Potential Consequences of Climate Variability and Change*, Report for the US Global Change Research Program (Cambridge, U.K.: Cambridge University Press, 2001); Committee on the Science of Climate Change, National Research Council, *Climate Change Science: An Analysis of Some Key Questions* (Washington, DC: 2001).

19. Houghton et al., op. cit. note 6, pp. 75–76.

20. Bert Metz, Ogunlade Davidson, Rob Swart, and Jiahua Pan, eds., *Climate Change 2001: Mitigation*, Contribution of Working Group III to the Third Assessment Report of the Intergovernmental Panel on Climate Change (Cambridge, U.K.: Cambridge University Press, 2001), pp. 3–5.

21. Ibid., pp. 5–8.

22. Ibid.

23. Ibid., p. 8.

24. Metz et al., op. cit. note 20, p. 9.

25. Ibid., p. 9; Luis Cifuentes et al., "Hidden Health Benefits of Greenhouse Gas Mitigation," *Science*, vol. 293, 17 August 2001, pp. 1257–59; Giulio A. DeLeo et al., "The Economic Benefits of the Kyoto Protocol," *Nature*, vol. 413, 4 October 2001, pp. 478–79.

26. Metz et al., op. cit. note 20, p. 10; Interlaboratory Working Group, *Scenarios for a Clean Energy Future* (Oak Ridge, TN: Oak Ridge National Laboratory and Berkeley, CA: Lawrence Berkeley National Laboratory, November 2000); European Climate Change Programme, European Climate Change Programme, Report—June 2001 (Brussels: June 2001); "European Union Says Enhanced Efficiency, Conservation Could Make Reductions Easier," *International Environment Reporter*, 20 June 2001, p. 513.

27. Metz et al., op. cit. note 20, p. 10.

28. Ibid.

29. Ibid.; Jae Edmonds, Joseph M. Roop, and Michael J. Scott, *Technology and the Economics of Climate Change Policy* (Arlington, VA: Pew Center on Global Climate Change, September 2000).

30. Stephen J. DeCanio et al., *New Directions in the Economics and Integrated Assessment of Global Climate Change* (Arlington, VA: Pew Center on Global Climate Change, October 2000).

31. Metz et al., op. cit. note 20, p. 11.

32. Ibid.

33. Ibid.

34. Ibid.

35. Ibid., pp. 11–12.

36. Metz et al., op. cit. note 20, p. 569; Organisation for Economic Co-operation and Development (OECD), *Environmentally Related Taxes in OECD Countries: Issues and Strategies* (Paris: October 2001); IEA in Metz et al., op. cit. note 20, p. 410.

37. Ibid., p. 12.

38. Ibid.

39. Ibid., pp. 12–13.

40. Figure 3 based on G. Marland, T.A. Boden, and R.J. Andres, Carbon Dioxide Information Analysis Center, Oak Ridge National Laboratory (ORNL), "Global, Regional, and National Annual CO_2 Emissions from Fossil-Fuel Burning, Cement Production, and Gas Flaring: 1751–1998 (revised July 2001)," at <cdiac.esd.ornl.gov/ndps/ndp030.html>, viewed 13 August 2001, and on BP, *BP Statistical Review of World Energy* (London: Group Media & Publications, 2001 and 2002); Figure 4 based on idem and on International Monetary Fund (IMF), *World Economic Outlook* (Washington, DC: April 2002); International Energy Agency (IEA), *CO_2 Emissions from Fuel Combustion, 1971-1999*, 2001 Edition (Paris: OECD/IEA, 2001), p. 101.

41. Text and Table 1 based on ibid.

42. Table 2 based on Marland, Boden, and Andres, op. cit. note 40, on BP, op. cit. note 40, and on Population Reference Bureau (PRB), *2001 World Population Data Set* (Washington, DC: 2002).

43. Data and Figure 5 based on ibid.; Protocol figure from Text of *Kyoto Protocol*, op. cit. note 2.

44. Data and Figure 6 based on Marland, Boden, and Andres, op. cit. note 40, on BP, op. cit. note 40, and on IMF, op. cit. note 40.

45. Zhou Dadi et al., "Australia," Report on the in-depth review of the second national communication of Australia, Submitted to the UN FCCC, 18 October 1999.

46. Ibid.

47. Ibid.

48. Ibid.

49. Ibid.

50. Ibid.

51. Text based on Marland, Boden, and Andres, op. cit. note 40, on BP, op. cit. note 40, and on PRB, op. cit. note 42.

52. Text and Figures 7 and 8 based on Marland, Boden, and Andres, op. cit. note 40, on BP, op. cit. note 40, and on IMF, op. cit. note 40.

53. Ministério da Ciência e Tecnologia, Governo da Brasil, "Convention on Climate Change: Brazil and the United Nations Framework Convention," at <http://www.mct.gov.br/clima/ingles>, viewed 3 May 2002; Eric Martinot, Akanksha Chaurey, Debra Lew, Jose Moreira, and Njeri Wamukonya, "Renewable Energy Markets in Developing Countries," in *Annual Review of Energy and the Environment*, vol. 27 (2002).

54. Ibid.

55. Ibid.; Martinot et al., op. cit. note 53.

56. Ibid.

57. Ibid.

58. Ibid.

59. Data and Figure 9 based on Marland, Boden, and Andres, op. cit. note

40, on BP, op. cit. note 40, and on PRB, op. cit. note 42; Protocol figure from Text of *Kyoto Protocol*, op. cit. note 2.

60. Data and Figure 10 are Worldwatch estimates based on Marland, Boden, and Andres, op. cit. note 40, on BP, op. cit. note 40, and on IMF, op. cit. note 40; Burhuan Nyenzi et al., "Canada," Report on the in-depth review of the second national communication of Canada, Submitted to the UN FCCC, 24 February 1999.

61. Nyenzi et al., op. cit. note 60.

62. Ibid.

63. Ibid.

64. Ibid.

65. Ibid.

66. Environment Canada, *Canada's Third National Report on Climate Change: Actions to Meet Commitments under the United Nations Framework Convention on Climate Change* (Ottawa: Minister of Public Works and Government Services, 2001).

67. Text based on Marland, Boden, and Andres, op. cit. note 40, on BP, op. cit. note 40, and on PRB, op. cit. note 42.

68. Text and Figure 11 based on Marland, Boden, and Andres, op. cit. note 40, and on BP, op. cit. note 40; Jonathan Sinton and David Fridley, "Hot Air and Cold Water: The Unexpected Fall in China's Energy Use," *China Environment Series*, Issue 4 (Washington, DC: Woodrow Wilson Center, 2001), pp. 3–20; John Pomfret, "Research Casts Doubt on China's Pollution Claims," *Washington Post,* 15 August 2001.

69. Jonathan E. Sinton and David G. Fridley, "Growth in China's Carbon Dioxide Emissions is Slower than Expected," *Sinosphere*, vol. 4, no. 1, winter 2001, pp. 3–5.

70. Ibid.; Jonathan E. Sinton and David G. Fridley, "What Goes Up: Recent Trends in China's Energy Consumption," *Energy Policy*, vol. 28, no. 10, August 2000, pp. 671–87.

71. Sinton and Fridley, op. cit. note 69; Figure 12 based on Marland, Boden, and Andres, op. cit. note 40, on BP, op. cit. note 40, and on IMF, op. cit. note 40.

72. ZhongXiang Zhang, "Is China Taking Actions to Limit its Greenhouse Gas Emissions? Past Evidence and Future Prospects," in José Goldemberg and Walter Reid, eds., *Promoting Development While Limiting Greenhouse Gas Emissions: Trends and Baselines* (New York: World Resources Institute (WRI)/U.N. Development Programme, 1999).

73. Sinton and Fridley, op. cit. note 68; Martinot et al., op. cit. note 53; Zhang, op. cit. note 72.

74. Text based on Marland, Boden, and Andres, op. cit. note 40, and on BP, op. cit. note 40; Lynch, op. cit. note 4; Protocol figure from Text of *Kyoto Protocol*, op. cit. note 2.

75. European Environment Agency (EEA), "EU Reaches CO_2 Stabilisation Target Despite Upturn in Greenhouse Gas Emissions," News Release, Copenhagen, 29 April 2002; Figures 13 and 14 based on Marland, Boden, and Andres, op. cit. note 40, on BP, op. cit. note 40, and on IMF, op. cit. note 40.

76. EEA, op. cit. note 75.

77. Ibid.; Worldwatch estimates based on Marland, Boden, and Andres, op. cit. note 40, and BP, op. cit. note 40.

78. Worldwatch estimates based on BP, op. cit. note 40.

79. Khaled Boukhelifa et al., "European Community," Report on the in-depth review of the second national communication of the European Community, submitted to the UN FCCC, 6 September 2000; IEA, *Dealing With Climate Change: Policies and Measures in IEA Countries* (Paris: OECD/IEA, 2000); Park Ill Soo et al., "Denmark," Report on the in-depth review of the second national communication of Denmark, Submitted to the UN FCCC, 2 October 1999; OECD, op. cit. note 36; Ministry of the Environment, Sweden, *Sweden's Third National Communication on Climate Change* (Stockholm: 2001).

80. Ibid; Worldwatch estimate based on BP, op. cit. note 40.

81. John Gummer and Robert Moreland, "European Union: A Review of Five National Programs," in Eileen Claussen, Vicky Arroyo Cochran, and Debra P. Davis, eds., *Climate Change: Science, Strategies, and Solutions* (Boston: Brill, 2001), pp. 88–115; Boukhelifa et al., op. cit. note 79.

82. Ibid.

83. Commission of the European Communities, *Third Communication from the European Community under the UN Framework Convention on Climate Change* (Brussels: 30 November 2001); Richard Rosenzweig, Matthew Varilek, and Josef Janssen, *The Emerging International Greenhouse Gas Market* (Washington, DC: Pew Center on Global Climate Change, March 2002).

84. Boukhelifa et al., op. cit. note 79; IEA, op. cit. note 40, p. xxviii.

85. Data based on Marland, Boden, and Andres, op. cit. note 40, and on BP, op. cit. note 40; Kyoto figure from Text of *Kyoto Protocol*, op. cit. note 2; Carlos Lopez et al., "Germany," Report on the in-depth review of the second national communication of Germany, Submitted to the UN FCCC, 24 August 1999.

86. Figure 15 based on Marland, Boden, and Andres, op. cit. note 40, and on BP, op. cit. note 40; Lopez et al., op. cit. note 85; Figure 16 based on Marland, Boden, and Andres, op. cit. note 40, on BP, op. cit. note 40, and on IMF, op. cit note 40.

87. Lopez et al., op. cit. note 85.

88. Ibid.

89. Ibid.

90. Ibid.

91. Ibid.

92. Ibid.; wind figure from Christopher Flavin, "Wind Energy Surges," in Worldwatch Institute, *Vital Signs 2002: The Trends That Are Shaping Our Future* (New York: W.W. Norton & Company, 2002), pp. 42–43.

93. Lopez et al., op. cit. note 85.

94. Data and Figure 17 based on Marland, Boden, and Andres, op. cit. note 40, and on BP, op. cit. note 40; 2000 rise based on EEA, op. cit. note 75.

95. Patricia Ramirez et al., "United Kingdom of Great Britain and Northern Ireland," Report on the in-depth review of the second national communication of the United Kingdom, Submitted to the UN FCCC, 17 December 1999; EEA, op. cit. note 75; Figure 18 based on Marland, Boden, and Andres, op. cit. note 40, on BP, op. cit. note 40, and on IMF, op. cit. note 40.

96. Ramirez et al., op. cit. note 95; Worldwatch estimates based on BP, op. cit. note 40.

97. Ramirez et al., op. cit. note 95.

98. Ibid.

99. Ibid.

100. Department for Environment, Food, and Rural Affairs, *The UK's Third National Communication under the United Nations Framework Convention on Climate Change* (London: 2001); U.K. Department for Environment, Food & Rural Affairs, "£215m Scheme Offers UK Firms Chance to Be World Leaders," Press release (London: 15 August 2001); idem, *Framework for the UK Emissions Trading Scheme* (London: August 2001); "Britain Leads the World in Tackling Climate Change," *The Economist*, 21 March 2002.

101. Ramirez et al., op. cit. note 95.

102. Data based on Marland, Boden, and Andres, op. cit. note 40, on BP, op.

cit. note 40, and on PRB, op. cit. note 42.

103. Data and Figure 19 based on Marland, Boden, and Andres, op. cit. note 40, and on BP, op. cit. note 40.

104. Figure 20 based on Marland, Boden, and Andres, op. cit. note 40, on BP, op. cit. note 40, and on IMF, op. cit. note 40; Rajendra Pachauri and Sudhir Sharma, "India's Achievements in Energy Efficiency and Reducing CO_2 Emissions," in Goldemberg and Reid, op. cit. note 72.

105. Pachauri and Sharma, op. cit. note 104; European Wind Energy Association (AWEA), American Wind Energy Association (AWEA), and Indian Wind Turbine Manufacturers Association (IWTMA), "Global Windpower Conference Heralds Major Clean Energy Expansion," Press Release, Paris, 3 April 2002; Metz et al., op. cit. note 20, p. 424.

106. Tata Energy Research Institute (TERI), "Climate Change Related Measures in India," at <http://www.teriin.org>, viewed 17 May 2002.

107. Pachuari and Sharma, op. cit. note 103.

108. TERI, op. cit note 106; EWEA, AWEA, and IWTMA, op. cit. note 105; "India Plans 6,000 MW Wind Power in Next 10 Years," *Reuters*, 3 April 2002.

109. Text and Figures 21 and 22 based on Marland, Boden, and Andres, op. cit. note 40, on BP, op. cit. note 40, and on IMF, op. cit. note 40.

110. Text of Kyoto Protocol, op. cit. note 2; French, op. cit. note 4; James Magezi-Akiki et al., "Japan," Report on the in-depth review of the second national communication of Japan, Submitted to the UN FCCC, 30 May 2000; Worldwatch estimate based on BP, op. cit. note 40.

111. Magezi-Akiki et al, op. cit. note 110.

112. Ibid.; Metz et al., op. cit. note 20, p. 418.

113. Ibid.; Magezi-Akiki et al., op. cit. note 110.

114. Ibid.

115. Ibid.

116. Ibid.; Molly O. Sheehan, "Solar Cell Use Rises Quickly," in Worldwatch Institute, op. cit. note 92, pp. 44–45; Magezi-Akiki et al., op. cit. note 110.

117. Magezi-Akiki et al., op. cit. note 110.

118. Ibid.

119. Data and Figure 23 based on Marland, Boden, and Andres, op. cit. note

40 and on BP, op. cit. note 40; Text of *Kyoto Protocol*, op. cit. note 2.

120. Nikitina Elena, "Russia: Climate Policy Formation and Implementation During the 1990s," *Climate Policy*, vol. 1 (2001), pp. 289–308; Figure 24 based on Marland, Boden, and Andres, op. cit. note 40, on BP, op. cit. note 40, and on IMF, op. cit. note 40.

121. Javier Hanna et al., "Russia," Report on the in-depth review of the second national communication of the Russian Federation, Submitted to the UN FCCC, 27 September 2000.

122. Ibid.

123. Ibid.

124. Ibid.

125. Ibid.; A. Mastepanov et al., "Post-Kyoto Energy Strategy of the Russian Federation, Outlooks and Prerequisites of the Kyoto Mechanisms Implementation in the Country," *Climate Policy*, vol. 1 (2001), pp. 125–33.

126. Data based on Marland, Boden, and Andres, op. cit. note 13, on BP, op. cit. note 40, and on PRB, op. cit. note 42.

127. Text and Figure 25 based on Marland, Boden, and Andres, op. cit. note 40, and on BP, op. cit. note 40.

128. Data and Figure 26 based on ibid. and on IMF, op. cit. note 40; Randall Spalding-Fecher, Energy and Development Research Centre, *Energy Sustainability Indicators for South Africa*, Prepared for Sustainable Energy and Climate Change Partnership (Cape Town: University of Cape Town, May 2002).

129. Spalding-Fecher, op. cit. note 128.

130. Ibid.; Martinot et al., op. cit. note 53; Department of Minerals and Energy, Republic of South Africa, *Draft White Paper on the Promotion of Renewable Energy and Clean Energy Development*, 21 June 2002, <http://www.dme.gov.za>, viewed 28 June 2002.

131. Ibid.

132. Data based on Marland, Boden, and Andres, op. cit. note 40, on BP, op. cit. note 40, and on PRB, op. cit. note 42.

133. Text and Figure 27 based on Marland, Boden, and Andres, op. cit. note 40, and on BP, op. cit. note 40.

134. Carlos Gay et al., "United States of America," Report on the in-depth review of the second national communication of the United States of America, Submitted to the UN FCCC, 12 May 1999; Figure 28 based on Marland,

Boden, and Andres, op. cit. note 40, on BP, op. cit. note 40, and on IMF, op. cit. note 40.

135. Henry Lee, Vicky Arroyo Cochran, and Manik Roy, "U.S. Domestic Climate Change Policy," *Climate Policy*, vol. 1 (2001), pp. 381–95; Gay et al., op. cit. note 134.

136. Gay et al., op. cit. note 134; U.S. Department of State, *Climate Action Report 2002* (Washington, DC: May 2002); Andrew C. Revkin, "U.S. Sees Problems in Climate Change," *New York Times*, 3 June 2002.

137. U.S. Department of State, op. cit. note 136.

138. Ibid.; Gay et al., op. cit. note 134.

139. U.S. Department of State, op. cit. note 136; Flavin, op. cit. note 92; U.S. Department of State, op. cit. note 136.

140. U.S. Department of State, op. cit. note 136; Michael Margolick and Doug Russell, *Corporate Greenhouse Gas Reduction Targets* (Arlington, VA: Pew Center on Global Climate Change, November 2001); Pew Center on Global Climate Change, *Climate Change Activities in the United States* (Washington, DC: June 2002); Jeffrey Ball, "California Opens New Front In Battle Over Fuel Efficiency," *Wall Street Journal*, 3 May 2002; John H. Cushman, Jr., "California Lawmakers Vote to Lower Auto Emissions," *New York Times*, 2 July 2002; Eileen Claussen, "The Global Warming Dropout," *New York Times*, 7 June 2002.

141. The White House, "Current Actions to Address Climate Change," 11 June 2001; The White House, "Global Climate Change Policy Book," 14 February 2002.

142. The White House, "Global Climate Change Policy Book," op. cit. note 141; Pew Center on Global Climate Change, op. cit. note 140; John J. Fialka, "Senate Committee Votes to Limit CO_2 Emissions from Power Plants," *Wall Street Journal*, 28 June 2002.

143. WRI, "Analysis of Bush Administration Greenhouse Gas Target," 14 February 2002; Pew Center on Global Climate Change, "Pew Center Analysis of President's Bush's February 14th Climate Change Plan," 14 February 2002; "Blowing Smoke," *The Economist*, 14 February 2002.

144. IEA, op. cit. note 79; Table 3 is a Worldwatch assessment using criteria and categories of idem, and based on this paper's case studies and research.

145. IEA, op. cit. note 79.

146. Ibid.; Metz et al., op. cit. note 20, p. 410.

147. IEA, op. cit. note 79.

148. Rosienzweig, Varilek, and Janssen, op. cit. note 83; trading estimate based on ibid. and on U.S. Environmental Protection Agency, *Latest Findings on National Air Quality: 2000 Status and Trends* (Washington, DC: September 2001); trading estimate based on Rosenzweig, Varilek, and Janssen, op. cit. note 83, and on Jack Cogen, Natsource LLC, "Opportunities and Obstacles Facing Emissions Trading Markets," Presentation to Transatlantic Dialogue on Climate Change Issues: New Ideas for a New Era, Center for Transatlantic Studies, Paul H. Nitze School for International and Area Studies, Johns Hopkins University, Washington, DC, 11 June 2002.

149. IEA, op. cit. note 79.

150. Ibid.; Metz et al., op. cit. note 20, p. 418; World Energy Council (WEC), "World Energy Council's Greenhouse Gas Emissions Reduction Projects Database Meets First Target," Press Release, London, 18 October 2001; U.N. Environment Programme and WEC, "Up to Two Billion Tons of Carbon Dioxide Saved by Cleaner Energy Schemes by 2005; Industry Acting to Fight Global Warming Despite Political Disagreements Over Kyoto," Press Release, Nairobi and London, 29 June 2001.

151. IEA, op. cit. note 79.

152. Metz et al., op. cit. note 20, pp. 11–12.

153. Worldwatch assessment based on ibid. and on above case studies.

154. Ibid.

155. Ibid.

156. Ibid.

157. Ibid.; p. 583.

158. Ministry of the Environment, Sweden, op. cit. note 79.

159. Daniel Sperling and Deborah Salon, *Transportation in Developing Countries: An Overview of Greenhouse Gas Reduction Strategies* (Washington, DC: Pew Center on Global Climate Change, May 2002).

160. Michael Grubb and Joanna Depledge, "The Seven Myths of Kyoto," *Climate Policy*, vol. 1 (2001), pp. 269–72; quote from Goldemberg and Reid, op. cit. note 72.

161. Worldwatch assessment based on Metz et al., op. cit. note 20, and on case studies above.

162. Joseph E. Aldy, Peter R. Orszag, and Joseph E. Stiglitz, "Climate Change: An Agenda for Global Collective Action," Prepared for Workshop on the Timing of Climate Change Policies, Pew Center on Global Climate Change, Arlington, VA, October 2001; Pew Center on Global Climate Change, "Nobel

Prize Recipient Dr. Joseph E. Stiglitz Calls for Immediate Action Against Climate Change," Press Release, Arlington, VA, 11 October 2001; William D. Nordhaus, "Global Warming Economics," *Science*, vol. 294, 9 November 2001, p. 1283.

163. Richard Schmalensee, "The Lessons of Kyoto," *Sloan Management Review*, vol. 43, no. 2, winter 2002, p. 96.

164. Council of Economic Advisors (CEA), Executive Office of the President, "Building Institutions for a Better Environment," Chapter 6, *Economic Report of the President*, Transmitted to the Congress (Washington, DC: Government Publishing Office, February 2002); Daniel Altman, "Just How Far Can Trading of Emissions Be Extended?" *New York Times*, 31 May 2002.

165. CEA, op. cit. note 164; "The Genie in the Wings," *The Economist*, 8–14 June 2002; R. Glenn Hubbard, Chairman, CEA, "The Way Forward on Climate Change Policy," Presentation to Transatlantic Dialogue, op. cit. note 148.

166. Edward A. Parson, "Moving Beyond the Kyoto Impasse," *New York Times*, 31 July 2001.

167. Rosenzweig, Varilek, and Janssen, op. cit. note 83.

168. Terry Dinan et al., *An Analysis of Cap-and-Trade Programs for Reduction of US Carbon Emissions* (Washington, DC: Congressional Budget Office, June 2001); Richard D. Morgenstern, "Reducing Carbon Emissions and Limiting Costs," Prepared for the Aspen Institute's Environmental Policy Forum, January 2002 (Washington, DC: Resources for the Future, February 2002); "Tax or Trade," *The Economist*, 14 February 2002; Daniel Bodansky, "Linking U.S. and International Climate Change Strategies," Working Paper, Pew Center on Global Climate Change, April 2002.

169. Michael Grubb, Jean-Charles Hourcade, and Sebastian Oberthur, *Keeping Kyoto: A Study of Approaches to Maintaining the Kyoto Protocol on Climate Change* (London: Climate Strategies, July 2001); Climate Strategies, "New International Research Organisation Argues that the Kyoto Protocol Can and Must be Rescued by the EU, Japan, and Russia—and That the US Will Then Come Back In," Press Release, London, 10 July 2001.

170. Box 3 consists of Worldwatch Institute recommendations.

171. Brian C. O'Neill and Michael Oppenheimer, "Dangerous Climate Impacts and the Kyoto Protocol," *Science*, vol. 296, 14 June 2002, pp. 1971–72.

172. Thomas Schelling, "What Makes Greenhouse Sense?" *Foreign Affairs*, vol. 81, no. 3, May–June 2002, pp. 2–9.

Index

Other Worldwatch Papers

On Climate Change, Energy, and Materials

157: Hydrogen Futures: Toward a Sustainable Energy System, 2001
151: Micropower: The Next Electrical Era, 2000
149: Paper Cuts: Recovering the Paper Landscape, 1999
144: Mind Over Matter: Recasting the Role of Materials in Our Lives, 1998
138: Rising Sun, Gathering Winds: Policies To Stabilize the Climate and Strengthen Economies, 1997
130: Climate of Hope: New Strategies for Stabilizing the World's Atmosphere, 1996
124: A Building Revolution: How Ecology and Health Concerns Are Transforming Construction, 1995

On Ecological and Human Health

153: Why Poison Ourselves: A Precautionary Approach to Synthetic Chemicals, 2000
148: Nature's Cornucopia: Our Stakes in Plant Diversity, 1999
145: Safeguarding the Health of Oceans, 1999
142: Rocking the Boat: Conserving Fisheries and Protecting Jobs, 1998
141: Losing Strands in the Web of Life: Vertebrate Declines and the Conservation of Biological Diversity, 1998
140: Taking a Stand: Cultivating a New Relationship With the World's Forests, 1998
129: Infecting Ourselves: How Environmental and Social Disruptions Trigger Disease, 1996

On Economics, Institutions, and Security

159: Traveling Light: New Paths for International Tourism, 2001
158: Unnatural Disasters, 2001
155: Still Waiting for the Jubilee: Pragmatic Solutions for the Third World Debt Crisis, 2001
152: Working for the Environment: A Growing Source of Jobs, 2000
146: Ending Violent Conflict, 1999
139: Investing in the Future: Harnessing Private Capital Flows for Environmentally Sustainable Development, 1998
137: Small Arms, Big Impact: The Next Challenge of Disarmament, 1997

On Food, Water, Population, and Urbanization

156: City Limits: Putting the Brakes on Sprawl, 2001
154: Deep Trouble: The Hidden Threat of Groundwater Pollution, 2000
150: Underfed and Overfed: The Global Epidemic of Malnutrition, 2000
147: Reinventing Cities for People and the Planet, 1999
136: The Agricultural Link: How Environmental Deterioration Could Disrupt Economic Progress, 1997
135: Recycling Organic Waste: From Urban Pollutant to Farm Resource, 1997
132: Dividing the Waters: Food Security, Ecosystem Health, and the New Politics of Scarcity, 1996

Other Publications From the Worldwatch Institute

State of the World Library
Subscribe to the *State of the World Library* and join thousands of decisionmakers and concerned citizens who stay current on emerging environmental issues. The *State of the World Library* includes Worldwatch's flagship annual, *State of the World*, plus all five of the highly readable, up-to-date, and authoritative *Worldwatch Papers* as they are published throughout the calendar year.

Signposts 2002
This CD-ROM provides instant, searchable access to over 965 pages of full text from the last two editions of *State of the World* and *Vital Signs*, comprehensive data sets going back as far as 50 years, and easy-to-understand graphs and tables. Fully indexed, *Signposts 2002* contains a powerful search engine for effortless search and retrieval. Plus, it is platform independent and fully compatible with all Windows (3.1 and up), Macintosh, and Unix/Linux operating systems.

State of the World 2002
Worldwatch's flagship annual is used by government officials, corporate planners, journalists, development specialists, professors, students, and concerned citizens in over 120 countries. Published in more than 20 different languages, it is one of the most widely used resources for analysis.

Vital Signs 2002
Written by Worldwatch's team of researchers, this annual provides comprehensive, user-friendly information on key trends and includes tables and graphs that help readers assess the developments that are changing their lives for better or for worse.

World Watch
This award-winning bimonthly magazine is internationally recognized for the clarity and comprehensiveness of its articles on global trends. Keep up to speed on the latest developments in population growth, climate change, species extinction, and the rise of new forms of human behavior and governance.